# Getting Into Area

## Hands-on Problem-Solving Activities for Grades 4–6

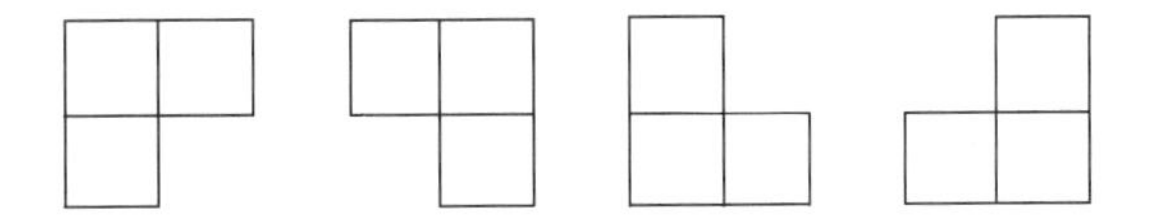

Anne M. Bloomer

Dale Seymour Publications

The point system for scoring written work on pages ix and 82 is adapted with permission from *Problem-Solving Experiences in Mathematics,* by Randall I. Charles and Frank K. Lester, © 1985 Addison-Wesley Publishing Company, Menlo Park, Calif.

The designs on pages 89–91 are reprinted with permission from *Tessellation Teaching Masters,* by Dale Seymour, © 1989 Dale Seymour Publications, Menlo Park, Calif.

**Project Editor:** Mali Apple
**Production:** Claire Flaherty
**Design Manager:** John F. Kelly
**Design:** Michelle Rose
**Cover Design:** Maryann C. Murphy

This book is published by Dale Seymour Publications, an imprint of the Alternative Publishing Group of Addison-Wesley Publishing Company.



Order number DS21337
ISBN 0-86651-833-9

4 5 6 7 8 9 10 – ML – 03 02 01 00 99

# Contents

*Italics type indicates the objectives for each lesson.*

# Introduction

In 1989 the National Council of Teachers of Mathematics published *Curriculum and Evaluation Standards for School Mathematics* to help define what students ought to know about mathematics. These standards, and those set forth in the *Professional Standards for Teaching Mathematics* (NCTM, 1991), are shaping mathematics education today. Educators are looking for ways to enable students to have and to feel they have mathematical power.

To this end teachers are trying to provide mathematics experiences for their students that promote discovery learning and problem solving, and that help students connect today's math lesson to other topics in mathematics, other curricular areas, and the real world. Tools rarely used two decades ago—such as cooperative grouping, technology, and opportunities for written and oral communication—are now used regularly in mathematics classes. *Getting Into Area* offers help to teachers of students in grades 3 through 6 who want to "teach to the standards" by providing discovery and problem-solving experiences in the areas of geometry and measurement.

## NCTM 5–8 Standards for Geometry

In grades 5–8, the mathematics curriculum should include the study of the geometry of one, two, and three dimensions in a variety of situations so that students can

- identify, describe, compare, and classify geometric figures
- visualize and represent geometric figures with special attention to developing spatial sense
- explore transformations of geometric figures
- represent and solve problems using geometric models
- understand and apply geometric properties and relationships
- develop an appreciation of geometry as a means of describing the physical world

Is it a question of just offering more, or do we really need to change *how* we're teaching? In 1957 Pierre van Hiele and Dina van Hiele-Geldof published the results of their research on how students learn geometry. In their definitive papers the van Hieles proposed that the understanding of geometry topics proceeds along certain sequential and hierarchical levels. Concepts implicitly understood at one level become explicitly understood at the next level, and each level has its own language and linguistic symbols, which makes communication and understanding difficult between levels. This has implications for elementary and middle school teachers, since it has been estimated that most high school sophomores are at Level 1, yet are expected to understand high school geometry courses that are often at Level 3.

## The van Hiele Levels of Geometry Understanding

**Level 0** Students recognize figures by their global appearance.
*This is basically a visual level. Students do not notice specifics.*

**Level 1** Students can analyze properties of figures, but do not interrelate figures or properties.
*At this level students see figures as collections of properties.*

**Level 2** Students can relate figures and properties, but do not organize sequences of statements to justify observations.
*At this level students can classify shapes into categories and can justify their choices but can't prove them.*

**Level 3** Students develop sequences of statements to deduce one statement from another, but do not recognize rigor or the relationships between other deductive systems.
*A typical geometry class is taught at this level.*

**Level 4** Students can analyze various deductive systems with high rigor. They understand properties of a deductive system with consistency, independence, and completeness of postulates.
*This is the level achieved by mathematicians.*

The van Hieles also suggested phases of learning that must occur for students to move to a higher level. This is the "how to help students" part of their model.

## The van Hiele Learning Phases

**Phase 1: Inquiry**

Teachers engage students in a two-way conversation about the objects of study.

**Phase 2: Directed orientation**

Teachers carefully sequence activities for student exploration. Students begin to realize what direction their study is taking. Students are starting to see relationships.

**Phase 3: Expliciting**

Students build from previous experiences with a minimum of teacher prompting.

**Phase 4: Free orientation**

Teachers provide multistep tasks or tasks with multiple ways to complete them. Students gain experience in finding their own ways to resolve tasks.

**Phase 5: Integration**

Teachers encourage students to review methods they know. Students form an overview of the subject under study. Objects and relations are unified and internalized.

It seems clear that there is a real need for students to work more with geometry during their elementary and junior high years, but that this work must be structured differently. Students need many hands-on experiences with both plane and solid figures. They must to be allowed and encouraged to construct, test, and modify their own ideas into a coherent whole. In this way they can develop understanding not only of concepts themselves but also of how they fit with other mathematical topics and with real life.

## Measurement

In looking at the teaching of geometry, we should not forget that many geometry topics are closely connected to ideas of measurement. This, too, is an area only touched upon by many teachers. Measurement chapters are among those that are often skipped in the press of time and concern with covering the curriculum. However, we do a disservice to our students in many ways when we concentrate on computation and arithmetic and only visit other topics quickly, if at all. If we skip stones over wave after wave of ideas, students never come to experience and appreciate the power and beauty of the waves themselves. They don't develop an appreciation for the myriad facets of mathematics, and they don't develop the feeling that they have mathematical power.

Because measurement chapters are often not taught, many teachers have not developed an understanding of how children learn measurement. As is true for the learning of geometry, students need to master a sequence of ideas to fully understand measurement topics. Patricia S. Wilson and Alan Osborne (1992) offer a particularly clear outline of the foundational ideas of measurement:

### The Foundational Ideas of Measurement

1. **Number Assignment**
   For every measurement there is a single number to represent that measurement. The number reports how many units in the measurement.

2. **Comparison**
   Like properties can be compared to see which is greater. The method of comparing properties varies from measurement system to measurement system.

3. **Congruence**
   Figures are congruent if they coincide when superimposed.

4. **Unit**
   There is a unit that can be assigned the number 1. The unit must be compatible with the property being measured.

5. **Additivity**

   Measurements of parts can be added to obtain the measurement of the whole.

6. **Iteration**

   A unit can be repeated to "cover" the property being measured. The number of iterations is the number assigned to that measurement.

Wilson and Osborne also suggest an instructional sequence that is based upon both research and the experience of many teachers:

### Suggested Instructional Measurement Sequence

1. **Number-free comparison**

   This helps students to focus attention on the attribute being measured rather than on the measurement tool. Example: Comparing areas to see which is bigger.

2. **Development and use of nonstandard units**

   This helps students begin to develop a numerical approach. Example: Measuring areas using index cards as the unit.

3. **Examination of the measurement system in terms of whether each foundational idea works within the system**

   This helps students see the structure of measurement itself and helps them focus on the similarities and differences between systems. Example: Asking students if iteration works or if units can be repeated to cover an area.

4. **Use of standard units**

   This develops students' abilities to understand the units in general use. Example: Measuring in square inches.

The early lessons in this book provide experiences to help students develop an understanding of measurement as it relates to area and perimeter. At the same time, students can begin to move forward through the van Hiele phases of learning. Later lessons focus more on the phases of learning as students use a variety of techniques to explore figures and concepts.

All experiences were developed with the NCTM curriculum geometry standards in mind. The activities are sequential, building on previously experienced concepts. If you choose to start in the middle, look at the bold print concepts in previous lessons to make sure your students have a firm foundation for new work. The table of contents clearly states the objectives covered in each lesson.

## Assessment

All teachers make continuous informal assessments of individual students and of the class's understanding as a whole. As you do more specific assessments, it is helpful to know *ahead of time* what you will be assessing. You may want to keep in mind the van Hiele phases of learning as well as the suggested instructional measurement sequence for an assessment of students' general levels.

You will also want to focus on the specific concepts being taught: How well do students seem to understand them? Can students use their understanding to solve problems? Can students use their understanding to help them see connections? At the developmental stage, it is more important to look at the process by which students arrive at an answer than at whether or not they calculate correctly.

The format of these lessons offers opportunities for many types of assessment: observations of students as they work, teacher-student and student-student communication, individual interviews, written responses, answers on student activity sheets, and journal writings. Journal writings can be especially helpful because they allow insight into students' attitudes, as well as into students' processes and understandings. If you wish to keep track of your observations and make a formal, scored assessment, you may want to use a rubric like that below, developed by Randall I. Charles and Frank K. Lester, Jr. (reproduced as a blackline master on page 82):

## A Point System for Scoring Written Work

### Understanding the Problem

**0** Completely misinterprets the problem.
**1** Misinterprets part of the problem.
**2** Completely understands the problem.

### Choosing and Implementing a Solution Strategy

**0** Makes no attempt or uses a totally inappropriate strategy.
**1** Chooses a partly correct strategy based on interpreting part of the problem correctly.
**2** Chooses a strategy that could lead to a correct solution if used without error.

### Getting the Answer

**0** Gets no answer or a wrong answer based on an inappropriate solution strategy.
**1** Makes copying error or computational error, gets partial answer for a problem with multiple answers, or labels answer incorrectly.
**2** Gets correct solution.

You may also want to develop a whole-class scoring sheet with students' names along the left-hand side and a space at the top to write the aspect being assessed (see page 83, which can be duplicated with the rubric on the back). Scores are marked in the column below this. Large graph paper is useful for this kind of recording, and the sheet itself facilitates looking at students' developing capabilities over a longer period of time. It is also useful in planning for instruction and an easy tool to use when making out grades or planning for conferences.

## A Last Piece of Advice

Do not be too concerned that these lessons take more time than you usually allow for geometry; students are learning lessons far and beyond what they would be learning by a "memorize the rule" approach, and far and beyond what they would be learning by looking at figures on the printed page for three straight days. And do not be too concerned about the noise level in your classroom. It will be the sound of students communicating, postulating, enjoying, thinking, and, above all, learning!

## References

Charles, Randall I., and Frank K. Lester, Jr. *Problem-Solving Experiences in Mathematics,* grades 1–8. (Menlo Park, Calif.: Addison-Wesley Publishing Company, 1985).

Fuys, D., D. Geddes, and R. Tischler. "English translations of selected writings of Dina van Hiele-Geldof and Pierre M. van Hiele." (New York: Brooklyn College, 1984.)

National Council of Teachers of Mathematics. *Curriculum and Evaluation Standards for School Mathematics.* (Reston, Va.: National Council of Teachers of Mathematics, 1989), p. 112.

National Council of Teachers of Mathematics. *Professional Standards for Teaching Mathematics.* (Reston, Va.: National Council of Teachers of Mathematics, 1989), p. 112.

Wilson, Patricia S., and Alan Osborne. Foundational Ideas in Teaching About Measure, ed. Thomas R. Post, *Teaching Mathematics in Grades K–8.* (Boston, Ma.: Allyn and Bacon, 1992), pp. 91–101.

# Lesson 1: Card Covers

### Measuring Area with Nonstandard Units

## Materials

4" by 6" index cards—several per student

half sheets of paper

masking tape

*optional:* 18" by 24" paper; chart paper

## Procedure

1. Pass out the index cards. Draw an 18" by 24" rectangle on the board, or tape up an 18" by 24" piece of paper. Ask students how big the rectangle or paper is. Explain: We want to cover the surface area with a collage. How much paper it will take to cover it? We'll measure with index cards.

2. Write student estimates on the board. One at a time, have students tape a card to the rectangle on the board. (*It takes 12 cards.*) Students may want to revise their estimates. Write: Area = 12 card units.

3. Start a list on chart paper or a bulletin board: "What We Know About Area." Ask: What can we say about area so far?
   ***Area is a measure of covering.***

4. Discuss other objects that have area.

5. Have students work in groups to measure the area of three objects in the classroom using their cards as a unit of measure. Have them make a table for their data with these headings:

| Object | Estimate of Area | Sketch of Cards | Area |
|---|---|---|---|

6. Discuss:
   - Did anyone need to estimate partial units?
   - How did you choose to handle that?
   - Who did it a different way?

7. List the objects and areas in a table on the board. Did more than one group measure the same object? Compare findings, and remind the students these measurements aren't very accurate, but are a starting place for finding area. Discuss results:
   - Does any object have the same area as another?
   - Are these objects the same size? Are they the same shape?
   - What had the largest area? The smallest?
   - How do you know? Could you know without measuring?
   - Did your estimates improve with practice? What new clues were you using?

8. Discuss what to add to the "What We Know About Area" list:

   ***We can add card units to get an area measurement.***

   ***Not all areas are whole numbers.***

   ***We can compare areas.***

   ***We can measure irregular areas.***

   ***Area is a two-dimensional measurement.***

   ***We need to measure with some unit that has its own area.***

## Journal Writing

*Complete this sentence:* Measuring area is like measuring length in the following ways:

*Complete this sentence:* Measuring area is not like measuring length in the following ways:

## Extensions

- Have students make a graph showing objects and their areas.
- Have students make an 18" by 24" collage.

# LESSON 2: SHAPING UP

### Measuring Area with Units of Different Sizes

## Materials

$3\frac{1}{2}$" by 7" sheets of construction paper—1 sheet per pair

pattern blocks (see pp. 84–87)—at least 8 hexagons, 16 trapezoids, 24 rhombuses, and 48 triangles per group

Shaping Up activity sheet—1 per student or pair

optional: isometric graph paper (see p. 88); markers

## Procedure

1. Have pairs use each pattern block in turn to complete the Shaping Up activity. Remind them to label results and to look for patterns or interesting findings.

2. Put a chart for findings on the board with these headings:

   **Block Type** **Measured Area**

   Students add their findings to the chart.

3. When all students have finished the activity, discuss the results. Students should see that the number of units it takes to cover the paper depends on the size of the unit; it takes more smaller units than larger units to measure an area. Students may see other relationships:
   1 hexagon = 2 red trapezoids = 3 blue rhombuses = 6 green triangles.
   Since a trapezoid is the same area as three triangles, the number for an area measured in triangles is three times larger than the number for trapezoids as a unit, etc.

4. Discuss:
   - What techniques did you use for estimating?
   - How did you find the area when blocks didn't cover the paper exactly?

5. Have students estimate, then use, a pattern block unit to measure one of the previously measured objects. Then have them compare their results with those they obtained from measuring with cards.

6. Discuss:
   - Was the estimating harder or easier? Why?
   - Do you think these measurements are less accurate or more accurate than the measurements using a card as a unit?
   - If you measured in hexagons, what would the measurement be in triangle units? (etc.)

7. Discuss new findings to be added to the "What We Know About Area" list:
   ***Use one kind of unit at a time for measuring.***
   ***A larger unit gives a smaller number for the measurement.***
   ***The number to describe the area depends on the size of both the area and the unit.***
   ***Smaller unit size means more accurate measurement.***
   ***If it takes two units to make a large unit, the area number when measured with the small unit will be twice as large as the area number when measured with the large unit.***
   ***If units are related, we can substitute one for the other.***

## Journal Writing

*Complete this sentence:* My partner and I worked well together when . . .
*Complete this sentence:* My favorite math discovery I made today is . . .

## Extensions

- Have students pick a pattern block to use as one unit. Have them find several ways to show an area of 15. They can color each design on isometric paper, cut out, and display.
- Relate pattern block relationships to fractions, especially equivalent fractions.

Name Date

## SHAPING UP

1. Use only one kind of block at a time. Estimate how many blocks it will take to cover the $3\frac{1}{2}$" by 7" sheet of construction paper.
2. Measure the area of your paper with the block you used.
3. Write your results in the chart. Label it! Example: 8 hexagon units.
4. Answer the questions. Use the back of this sheet if you need more room.

| Block Type | Estimated Area | Measured Area |
|---|---|---|
| Hexagon | | |
| Trapezoid | | |
| Rhombus | | |
| Triangle | | |

I see these patterns on the chart:

Discuss with your partner what you can say about area. Have you made any new discoveries?

# Lesson 3: Marvelous Mosaics

### Exploring Areas of Congruent Figures and Halves of Symmetrical Figures

## Materials

crayons or markers in pattern block colors

isometric graph paper (see p. 88)—several sheets per student

pattern blocks (see pp. 84–87)—a few hexagons, trapezoids, rhombuses, and triangles per group

transparencies of tessellation patterns (mosaic designs) (see pp. 89–91)

scissors

## Procedure

1. Show the transparencies of mosaic designs. Have students identify the design unit that repeats. Ask: Would knowing the area of a design help in designing artwork or architecture?

2. Have students choose one type of pattern block, place several of them on the graph paper to make a design, determine its area in triangle units, trace around the design, and cut it out. Then have them place the cut-out design on graph paper, trace around it, and determine the area of the tracing.

3. Ask: Is your design and its tracing the same size? The same shape? Tell students that figures are called *congruent* if they have the same shape and size. Discuss what they discovered about the areas of the two congruent figures. Add to the "What We Know About Area" list: ***Figures that are congruent have equal areas.***

4. Have student groups develop a set of ten pattern blocks to use for a design. Each student in the group will then assemble the set, create a design with it, and find the area of the design. (The area will be identical for all designs within a group.)

5. Have groups compare designs and areas within and between groups. Discuss:
   - Are the designs congruent?
   - Are the areas the same for each design? How do you know?
   - Do figures or objects have to be congruent to have the same area?

   Add to the "What We Know About Area" list:
   ***Figures do not have to be congruent to have the same area.***

6. Have students next make a small pattern block design with one straight edge. This design will be half of a larger, symmetrical design. They copy this design onto isometric paper, color it, and determine its area. Then they color in its symmetrical half (across the straight side), find its area, and cut out both halves.

7. By rotating one of the halves 180°, students can compare the two halves. Discuss:
   - Are your two designs halves of a symmetrical design?
   - What did you discover about the areas of the two halves?
   - Are the halves congruent? How does this fit with our knowledge of area?
   - Are the halves identical? If there are no non-identical halves, have students consider a line of symmetry for the letter M. The halves are congruent if you allow flips.

   Add to the "What We Know About Area" list:
   ***Symmetrical halves have equal areas.***

## Journal Writing

Write a letter to an imaginary pen pal explaining what you learned about area today.

## Extensions

- Have students outline, on isometric paper, a rectangle 4 or 5 hexagons wide and 2 hexagons high. This will be a mosaic with repeating design units; they must have at least one repetition of their design within the rectangle. If they want to repeat their design twice, how much area must the design cover? What if they want to repeat their design 4 times? Have them use pattern blocks and markers to finish their designs. Cut out and post.
- Have students use eggshell pieces or seeds to cover mosaic designs.
- Using place value blocks, have students make a particular number several different ways; for example, the number 21 made with 2 tens and 1 one, 21 ones, or 1 ten and 11 ones. Ask: Are the areas the same? Connect this experience to regrouping.

# Lesson 4: Geosquares

**Relating Number and Area Patterns**
**Understanding Square Numbers**

## Materials

tiles—9 per student

geoboards and rubber bands, or geoboard paper (see p. 92)—1 per student

1-cm graph paper (see p. 96)—1 sheet per student

markers in different colors

Geosquares activity sheet—1 per student

## Procedure

*Note:* This lesson assumes use of a 5 by 5 geoboard. If a 6 by 6 geoboard is used, an extra square can be found.

1. Ask students to name some of the patterns they have found in mathematics. Explain that they will be searching for more patterns using geometric shapes.

2. Have students use geoboards and rubber bands to complete the Geosquares activity sheet.

3. Discuss:
   - What patterns did you find?
   - How are the squares the same? Different?
   - Are squares rectangles?
   - How are squares similar to and different from other rectangles?
   - How did the area of the squares change as the length of the sides increased?

   Students should see that the lengths of the sides are the same for squares. Relate the lengths of the sides to the factors in multiplication: 5 rows of 5 square units each means 25 square units in all; $5 \times 5 = 25$. Some students may see that the area increases more than they might expect.

4. Have students use tiles to form a square (if possible) from each number of tiles, from one tile to nine tiles. What do they notice? Discuss:
   - Not all numbers of tiles can be made into squares.
   - The numbers that can be made into squares (*2, 4, 9*) are numbers that can be made by multiplying two other identical numbers together.

   Explain that these numbers are called *square numbers.* If you wish, explain that a 2 by 2 square has an area of 4, which can be written $2^2$.

5. Have students use markers and graph paper turned horizontally. They will make the square numbers 1, 4, 9, 16, and 25 in the following way: for the number 1, they color in one space and write 1 below it. For the next number, 4, they start by coloring one square using the same color as before. This beginning square will always be in the lower left-hand corner of any new square. Changing colors, they color the necessary spaces to make a 2 by 2 square, and write 4 below it.

   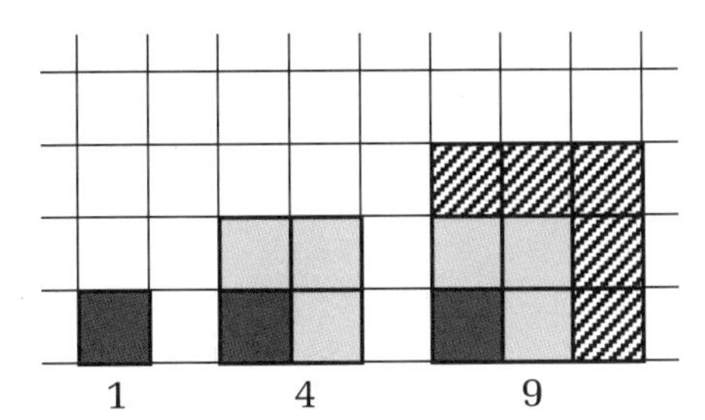
   

   New numbers follow this pattern, adding to the right-hand side and top of the previous number and using a different color to show the additions. When students have finished, have them write on the graph paper a rule for finding successive square numbers. (*Add 1 to each previous factor and multiply the new numbers together.*)

6. Discuss their rules and any other discoveries they have made. Ask:
   - How would you get to the next square number?
   - What will it be?
   - How many square numbers do you know?

7. Explore the relationship of the length of the sides to the area. First have students turn over their graph paper. Explain: A piece of jewelry is shown in a catalog advertisement. The picture in the advertisement measures 1" by 1". The newspaper says that the jewelry is actually three times bigger. What does this mean? What is its area?

   If the piece actually measures 3" by 3", how much bigger is the real area than the 1 square inches shown in the picture? Many students will feel that the area will be 3 square inches. Have students turn over their graph paper and check. Repeat, having students suggest other scenarios.

   Add to the "What We Know About Area" list:

   ***Square numbers are those whose factors are the same.***

   ***Square numbers only have one shape—a square.***

   ***To get to the next square number, add another column and another row.***

   ***Doubling the side lengths of a square more than doubles the area.***

## Journal Writing

Imagine that you are planning a seating chart for a class and you would like the desks to make a filled-in square. Is it possible for your class? How would knowing about square numbers come in handy? Give examples.

## Extensions

- Have students make a bar graph of the square numbers, using the length of the sides along the $x$-axis (horizontal) and the area along the $y$-axis (vertical). Students can see how quickly numbers grow when they are squared.

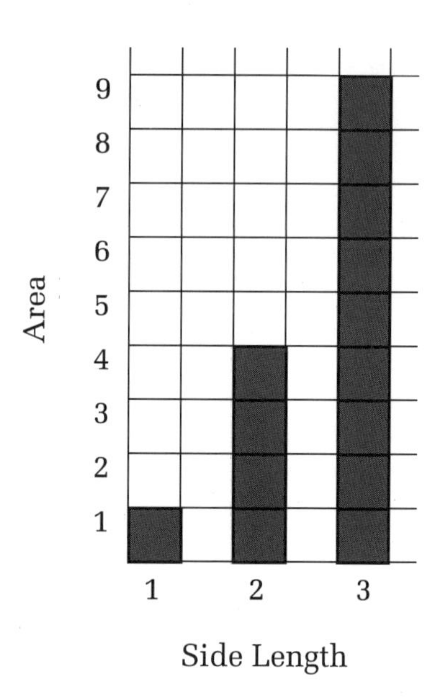


- Have students bring in catalog advertisements and repeat step 7 using actual measurements.
- Ask: If you add two square numbers, will you have another square number? Why or why not? What numbers can be made from other square numbers? For example, 16 can be made from four 4s.

Name Date

# GEOSQUARES

1. On your geoboard, make four squares that follow these rules:
   - Each square must have a different area.
   - All sides must be parallel with a geoboard edge. No slanted lines allowed.
2. Draw each square you made, fill in the data, and write your observations.

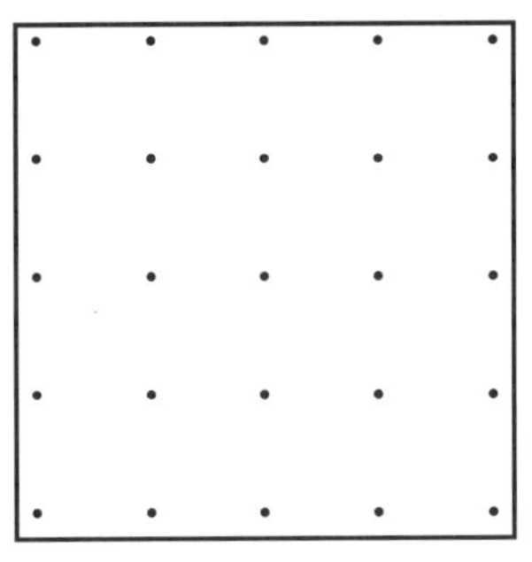

Side = _____ units

Side = _____ units

Area = _____ sq. units

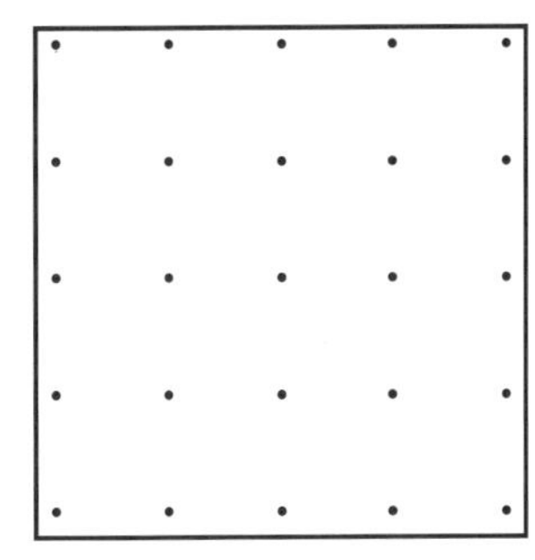

Side = _____ units

Side = _____ units

Area = _____ sq. units

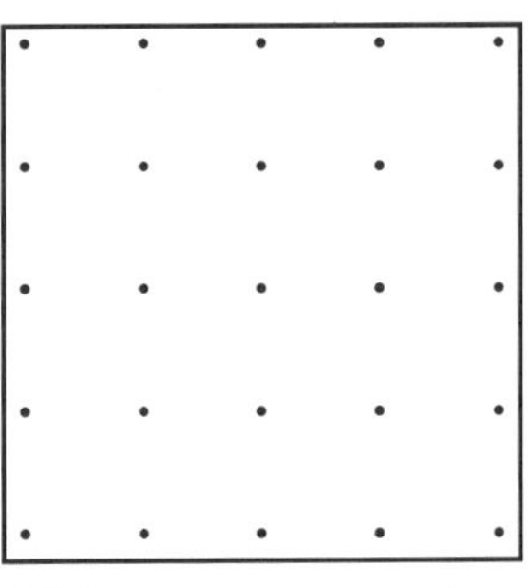

Side = _____ units

Side = _____ units

Area = _____ sq. units

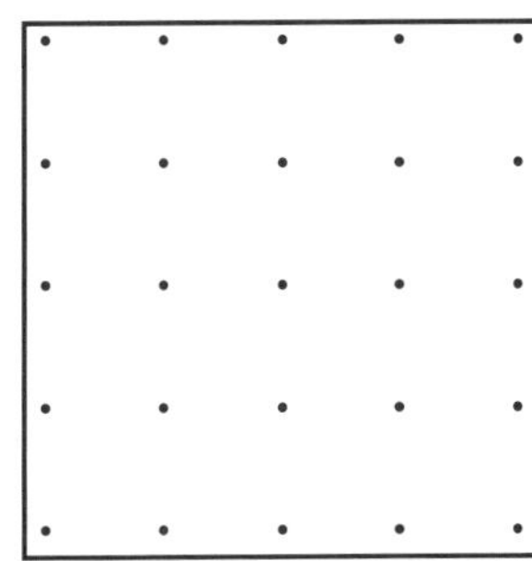

Side = _____ units

Side = _____ units

Area = _____ sq. units

What do notice about the lengths of the sides?

What do you notice about the areas?

# Lesson 5: Garden by the Numbers

**Exploring Area of Figures of More Than One Rectangle**
**Measuring Area with Standard Units**

## Materials

dice—1 per pair

geoboards and rubber bands, or geoboard paper (see p. 92)—1 per student

1-cm graph paper (see p. 96)—1 sheet per student

tiles—12 per pair

meter sticks

Garden by the Numbers activity sheet—1 per student or pair

## Procedure

1. Have students use rubber bands to enclose a 2 by 3 rectangle on their geoboards. They then outline a 1 by 1 square in a different area of the geoboard. They find the area of each figure, using a 1 by 1 square as 1 square unit. Next have them move the single square to adjoin the rectangle and put a rubber band around both areas.

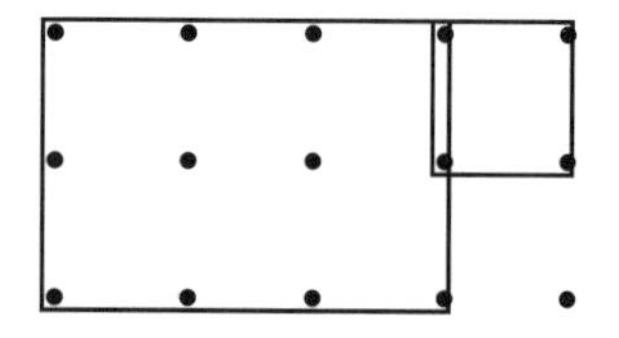

   Discuss:

   - What was the area of each figure? (*6 square units, 1 square unit*)
   - What is the area of the new figure? (*7 square units*)
   - How could we write this new area as an equation? (*6 + 1 = 7 square units*)
   - If we look at the figure as one large rectangle with a piece missing, how could we write an equation for the area? (*8 – 1 = 7 square units*)

2. Repeat several times with different sizes of rectangles, asking the same questions.

3. Have students make a figure on the geoboard with an area of 8 square units, copy it onto graph paper, and label the area. Then have them use another rubber band to separate the geoboard figure into two parts. They mark the same division on their graph paper rectangle and label the areas. Have them write associated addition and subtraction equations.

4 (width), 2 (height)

| 1 | | 3 | |
|---|---|---|---|
| A | | B | |

Area of large rectangle: $6 + 2 = 8$

Area of rectangle A: $8 - 6 = 2$

Area of rectangle B: $8 - 1 = 6$

4. Have students divide the figure into three parts, draw the parts, and write the appropriate addition and subtraction equations. Discuss:
   - What equations are possible with an area of 8?
   - Does this remind you of any other math topic? (*addition and subtraction facts*)

5. Discuss what they have discovered. Add to the "What We Know About Area" list:
   ***Small areas in a larger area can be added to find the large area.***
   ***Parts of an area can be subtracted, and equations can be written to show this.***

6. Discuss possible real-life problems, such as spilling water on 5 tiles of a floor covered with 40 carpet tiles. Using a tile as 1 unit, how much of the area is dry? (*35 carpet-tile units*)

7. Have pairs use dice and tiles to work on the Garden by the Numbers activity. Have several pairs sketch one of their gardens on the board. Discuss:
   - Did everyone who had an area of 4 use the same shape? What about the other areas?
   - Which gardens had more possible shapes? Why?
   - What kinds of things might you have to consider when planning a garden?

8. Have students look at one of their gardens. Ask: Can you can tell how big your garden is in real life? (*not without knowing the size of the unit*) Introduce square meters, and label the sketches on the board with meters. Explain: The units are called *square meters* because the squares are 1 meter in length on each side. Use meter sticks to lay out one of the gardens on the floor.

## Journal Writing

Interview an adult and ask how they use a knowledge of area in their life and work. Write a paragraph in your journal about the interview.

## Extensions

- Have groups create multisided gardens on graph paper. Sides should not be diagonal. More advanced students might include spaces for paths. Students estimate the area of the garden, label the sides, then find the area. Which kinds of shapes are harder to estimate?
- Get seed catalogues and have students plan gardens—$\frac{1}{2}$ beans, $\frac{1}{4}$ onions, etc. How far apart will they need to plant their plants? How many will they have of each type of plant?
- Have students use place value blocks to make gardens with areas of 10, 12, and 16 square feet. Which numbers give the largest variety of shapes? How many ways can they make a garden with an area of 36 square feet? Relate this activity to multiplication and fractions.

Name Date

# GARDEN BY THE NUMBERS

Darren and Shana can't decide on the shape of their garden this year. They know they have 9 cups of fertilizer left from last year. They use 1 cup of fertilizer for each square unit of garden. Will they have enough for each garden? Will they have some left over?

1. Design three gardens for Darren and Shana. For each garden, roll two numbers. Add them together to find the total area for that garden. Example: If you first roll a 2 and a 4, the area for Garden 1 will be 6 square units.
2. Use that many tiles to create a garden. Each garden's shape must be made of two or more rectangles joined together. Sketch each garden below. Label the lengths of the sides. For example, your garden might be:

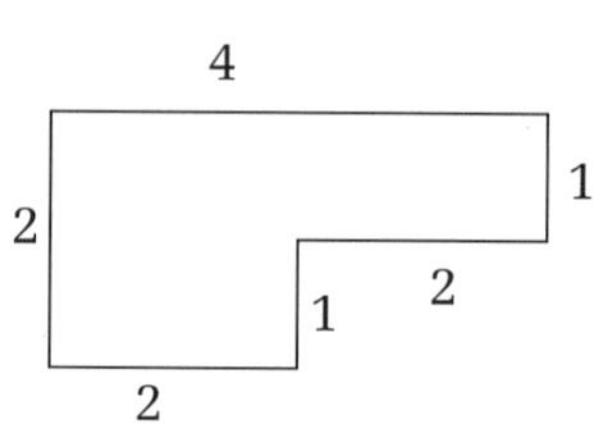


3. Draw lines to show the rectangles that make up each garden. Write an addition and a subtraction equation for each garden. Example:

4
Area = 4 1
2
Area = 2 1 2
2

$2 + 4 = 6$ square units

$8 - 2 = 6$ square units

4. Tell how much extra fertilizer Darren and Shana have or need to buy.

| Labeled Sketch | Equations | How Much Extra Fertilizer? How Much More Needed? |
|---|---|---|
| Garden 1 | | |
| Garden 2 | | |
| Garden 3 | | |

5. Choose one of the areas you used. On the back of this sheet, sketch as many different shapes for that area as you can. Which gardens have more possible shapes? Why?

# Lesson 6: Brush Up on Painting

**Developing a Formula for the Area of a Rectangle**
**Figuring Surface Area of Solids**

## Materials

calculators

cubes, preferably interlocking—16 per pair

$\frac{1}{4}$-in. graph paper (see p. 93)—1 sheet per pair

glue stick

pictures of buildings

scissors

straightedge—1 per pair

tiles—12 per pair

Rectangles, Rectangles activity sheet—1 per student or pair

Brush Up on Painting activity sheet—1 per student or pair

*optional:* exterior blueprints; half-pint milk cartons

## Procedure

1. Have students make a rectangle three tiles long (horizontally) and one tile wide. Start a chart with these headings:

   | Object | Length ($L$) | Width ($W$) | Area |
   |---|---|---|---|

   Ask: What is the area? (*3 square units*) Have them imagine that the tiles are stretchy material such as that used to make balloons. If they were to stretch the rectangle so it remained three tiles long, but was two rows wide instead of one, what would the area be? (*6 square units*) Ask: What strategies did you use to make your predictions? Have them add additional tiles to check.

2. Repeat several times, having students add another row of three tiles each time. Ask: Do you see a pattern on the chart?

3. Have students make a rectangle five squares long and one square wide. Ask: What will the area be if the material is stretched to be four tiles wide while remaining five tiles long? (*20 square units*) Discuss strategies. Some students may already see they can multiply; many will use serial addition. Discuss:
   - How many rows of five will you have?
   - How much is four rows of five?
   - Is this anything like something you've done before?

   Relate this concept to the multiplication they have done before: the tiles form an array, and the numbers on the chart make multiplication facts. Multiplication can be thought of as a shortcut for serial addition.

4. Have pairs use 12 tiles, glue, scissors, and graph paper to do the Rectangles, Rectangles activity.

5. Discuss:
   - How did you choose to order your charts?
   - What patterns do you see?
   - Did you find a way to find area?
   - Does multiplying the length times the height work for all the rectangles?

6. Have students help you make a general equation that works for the chart:
   Area = Length × Width
   Have students write this equation and then create some problems, such as finding the area of a farm 6 miles long by 2 miles wide. They write the numbers in equation form below the words: Area = 6 × 2. Discuss the need to state the product in square miles.

7. When students are comfortable with this substitution, introduce the use of letters *L* and *W*: $A = L \times W$. Discuss that a *formula* is a math language shortcut to describe a pattern or relationship. Add to the "What We Know About Area" list:
   ***Find rectangular area by multiplying length by width.***
   ***The formula for finding the area of a rectangle is*** $A = L \times W$.

8. Show building pictures and blueprints. Have students discuss the flat surfaces that could be measured, including the roof. Then have students use cubes to build their buildings for the Brush Up on Painting activity.

9. Discuss:
   - How much area needs to be painted?
   - Are some levels of the buildings smaller than others?
   - What about the roofs?

10. Remind students that they will have to change the building dimensions to feet. Have students use calculators to complete page 1 of the Brush Up on Painting activity sheet.

11. Have several students share their charts. Work through the directions for page 2 with students, then have them complete the page. At the conclusion, have several students show their buildings and report on the cost of their paint.

## Journal Writing

*Complete these sentences:* The new formula I learned is ________ . It means . . .

*Complete these sentences:* The activity that helped me understand this way of finding area was ________ . It helped me because . . .

## Extensions

- Have students collect half-pint milk cartons from their school lunches. They can cut and tape the cartons to create buildings, complete with doors, windows, and roofs. They can then decide on a simple scale, measure each flat surface, and use the Brush Up on Painting activity sheet to determine the amount of paint that will be needed to paint the buildings.
- Have students draw the buildings they created in the Brush Up on Painting activity. They can color the buildings and indicate the amount and cost of the paint needed to paint the actual buildings.
- Have a painting contractor or architect visit the class.

Name Date

## RECTANGLES, RECTANGLES

1. Make all the 12-cube rectangles that are possible. Copy each rectangle onto graph paper twice.
2. Cut out all rectangles. Rotate one of each pair. Example: I can make a rectangle 2 tiles long and 6 tiles wide. I cut out two of these and rotate one so that it is 6 tiles long and 2 tiles wide.
3. Glue the rectangles onto the chart below. Make sure 1 of each pair is rotated. Label the sides. Write the area on each.
4. Fill in the chart. Use your straightedge to draw lines between the rows. Look for a pattern in your chart that will help you find area.

| Rectangles | Length (horizontal) | Width (vertical) | Area |
|---|---|---|---|
| | | | |

5. What did you discover about finding area?

Name Date

## BRUSH UP ON PAINTING, PAGE 1

1. Use the cubes to build a simple building. Sketch your building below.

2. Imagine that you are going to paint your building. To know how much paint to buy, you will need to know the area in square feet of each flat surface. Let each cube width stand for 10 feet.

3. Decide: Will you use more than one color? Will you have doors and windows? Does the roof need painting?

4. Draw on the chart below all flat areas of your building. Label the sides of each flat area *in feet* and fill in the rest of chart.

| **Flat Surfaces (draw each one)** | **Length** $L$ | **Width** $W$ | **Area** $A = L \times W$ | **Color to Be Painted** |
|---|---|---|---|---|
| | | | | |

Name Date

## BRUSH UP ON PAINTING, PAGE 2

5. In the chart, list the colors and all the areas that will be painted that color. Add the areas to get a total.

   Example: My house's walls will be red. The areas are 200 sq. ft., 200 sq. ft., 100 sq. ft., and 100 sq. ft. Total red = 600 sq. ft.

6. Each gallon of paint covers 200 square feet. Find how many 200 square-foot areas you have for each color. Then decide how many gallons you need of that color.

   Example: I have 600 square feet to paint red. How many gallons will I need?

7. Do each color in turn, showing your work for each.

Figuring How Much Paint Will Be Needed

| Color | Areas to Be Painted That Color | Amount of Total for That Color | Paint Needed |
|---|---|---|---|
| | | | |

Finding How Much the Paint Will Cost

| Color | Gallons Needed × Cost per Gallon ($20) | Cost for This Color |
|---|---|---|
| | Total Cost: | |

# Lesson 7: Be Your Own Architect

## Finding the Area of Figures Made of More Than One Rectangle

### Materials

$\frac{1}{2}$-in. or 1-cm graph paper (see pp. 94 and 96)—several sheets per student

place value blocks: ones, tens, hundreds

plain paper

transparency of floor plans in steps 2 and 3 (see p. 97)

optional: architect-drawn floor plans; *Consumer Reports* article on interior paint (e.g. May 1991, p. 335)

### Procedure

1. Find out the minimum square footage of classroom floor mandated by the state. Have students use place value blocks to design classrooms with that much area. Let each side of a one cube equal 1 foot. Then have students transfer the designs to graph paper, labeling all dimensions in feet.

2. Draw the following floor plan on the board or show it on the overhead:

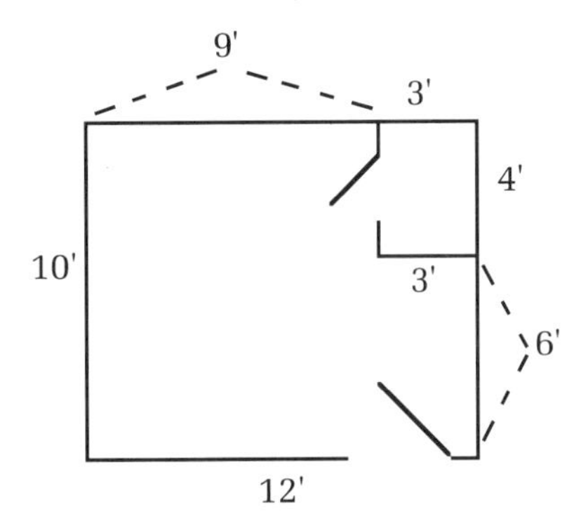


Say: The closet will not get carpeted. How much square footage will need to be carpeted? Discuss the strategies students use. Show architects' floor plans if you have them, and point out measurements. Discuss the various aspects of a room—doorway width, electrical outlets, windows, heating concerns—architects must consider.

3. Change the numbers on the floor plan as shown below, or show the floor plans on the overhead. Ask: How much square footage will need to be carpeted? Students will need to find the missing dimension before figuring the area. Discuss their strategies.

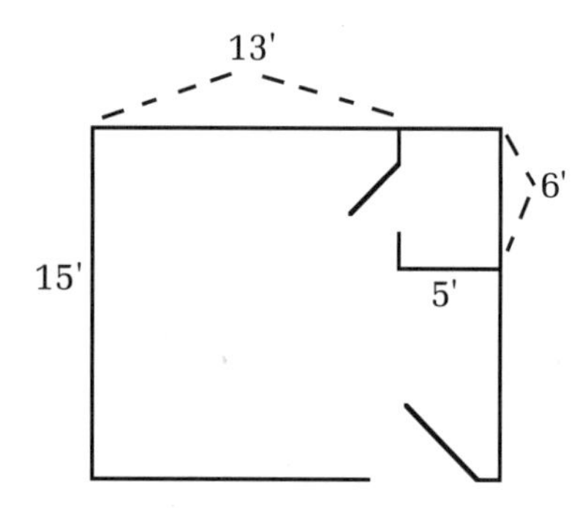


4. Have pairs use graph paper to design a floor plan with more than one rectangle, mark all dimensions, and figure the area. Then they transfer the design to plain paper and label with just enough measurements so someone else could figure out the dimensions. Have pairs exchange problems, solve, then check, using the original graph-paper floor plan.

5. Carpeting is measured in square yards, not square feet. Have students draw a 6 by 9 rectangle on graph paper, label it 2 yards by 3 yards (each square equals 1 square foot), and find the area. (*6 square yards*) Have students predict what the area will be in square feet. Then have them change the measurements to 6 feet by 9 feet and find the square footage.

   Most students will have assumed the area will be three times—not nine times—larger. Students can count the squares in their rectangle. Discuss this issues, relating it to students' previous experience with making square numbers. Repeat with other rectangles.

   Add to the "What We Know About Area" list:
   ***We can find area of irregular figures by looking at their parts.***
   ***When we change units of measurement for the sides of a rectangle, the area will reflect the changes made in*** **both** ***the length and width of the rectangle.***

## Journal Writing

Finish this sentence: Today I used my thinking ability in math when I . . .

## Extensions

- Have students estimate or actually measure their rooms, draw a floor plan, and estimate the square yardage of carpet required. If carpet is $15 per yard, what will the total cost be?

- For older students, have a carpet salesperson visit and explain how to figure yardage. Carpet comes in 12-foot (sometimes 15-foot) widths, which makes calculations more difficult. Would carpet tiles be easier or cheaper to work with? They, too, come in standard dimensions. What are the advantages of each?
- Have students find the area of the walls for painting their rooms, using a ceiling height of 8 feet. They will need to measure the windows and doors and subtract them. The area covered by a coat of paint varies with brand and type of paint. *Consumer Reports* tests paint for coverage—some paints are less expensive, but will require two coats to hide previous colors. Some paints resist scrubbing better than others. Have students choose a brand and color and justify their choices. If you can't find actual data, use these examples:

| **Brand** | **Cost/Gal.** | **Hiding Power** | | **Scrubbing** |
|---|---|---|---|---|
| | | ***1 coat*** | ***2 coats*** | |
| Sal's Superior–white | $31 | poor | good | good |
| Sal's Superior–pink | $31 | good | excellent | good |
| Sal's Superior–blue | $31 | good | excellent | good |
| Apple Advantage–white | $25 | poor | good | good |
| Apple Advantage–pink | $25 | poor | good | good |
| Apple Advantage–blue | $25 | poor | good | good |
| Kane's Casual Living–white | $15 | poor | poor | excellent |
| Kane's Casual Living–pink | $15 | poor | good | excellent |
| Kane's Casual Living–blue | $15 | poor | good | excellent |

- Have students explore floor plans with fractional dimensions (e.g. $12\frac{1}{4}$ feet) or decimal dimensions (e.g. 1.5 feet). This is a good chance for students to explore multiplication of fractions and decimals informally.

# Lesson 8: Mondrian Madness

**Using Rectangular Area Knowledge for Art**

## Materials

9" by 12" white construction paper—1 sheet per student

9" by 12" construction paper in different colors—1 sheet per student

paintings (art prints or photographs) by Piet Mondrian

large paper bag

paper cutter

glue sticks

rulers—1 per student

*optional:* calculators; black wide-line markers; larger sheets of black construction paper (for frames); tape; salad or mineral oil and paper towels (to create a stained glass effect)

## Procedure

1. Show the Mondrian paintings and discuss shapes, balance, and color.

2. Have students watch as you prepare the materials: mix up the stack of colored paper, then use a paper cutter to cut the colored paper into ten rectangles as shown. Draw the cutting pattern on the board if you wish. Make sure students know there is a sheet of paper for each of them.

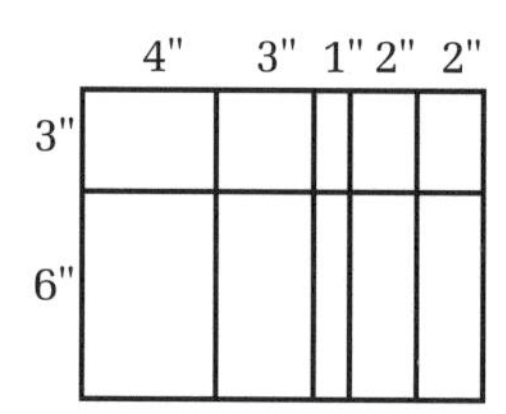


3. Put all the pieces into the bag and mix them. Pass out one white sheet and a handful of colored rectangles to each student (students do not all have to have the same number of pieces).

4. Have students measure the dimensions of each piece, mark the dimensions along the length and width of each piece, and note the area in the center. Then have them find the area of the white paper and the total area of their colored pieces. Discuss the areas of colored paper the students have. (They will vary.)

5. Students will be covering the white paper with colored rectangles to make a Mondrian-type collage with no uncovered white space and no overlapping rectangles. They will need to figure their missing or extra area, find enough pieces to fit, and negotiate trades for colors and sizes. Example: I am missing 12 square inches of area. What pieces could I use? Which size will fit with what I already have? Could I use two pieces? Would it help if I first arranged the pieces I already have?

6. Students find and trade pieces until they have an arrangement that pleases them, then glue the pieces onto the white paper. If you wish, have students outline the boundaries of the rectangles with black marker, then cut frames from black paper and tape their art behind the frames. They can also spread oil on their pictures, wiping off the excess. (Note: Oil dissolves some inks. Test the construction paper before trying this.) Hang the art in a window and the light will shine through, giving a stained glass effect.

## Journal Writing

Write about five ways in which you could use your knowledge of area as an artist or architect.

## Extensions

- Show Escher drawings and discuss how his knowledge of area was helpful in his artwork.
- Introduce tiling (tessellating) with pattern blocks. A shape tessellates a plane if it completely fills it with no uncovered spaces. What polygons tessellate? What combinations of polygons tessellate? (See pp. 89–91 for tessellation masters.)
- Have students create flags or designs on graph paper. Specify fractional amounts such as $\frac{1}{4}$ red, $\frac{5}{8}$ blue, and so on. How many squares will be colored each color?

# Lesson 9: Inchworm Investigations

**Measuring Perimeter**

**Relating Perimeter and Area**

## Materials

1 meter length of string—1 per group

meter sticks—1 per group

1" tiles—4 per student

String Strategies activity sheet—1 per group

Inchworm Investigations activity sheet—1 per student

*optional:* chart paper

## Procedure

1. Have groups of students measure the distance around the edge of the classroom using any method and unit of measure they choose. Have them put their results on the board under the headings: Distance Around; Unit of Measure. Discuss that they have been measuring *perimeter,* the distance around something. Ask: How can we compare these perimeters? Students should see that identical units of measure are necessary for any comparison.

2. Have groups use string and meter sticks to complete the String Strategies activity and sheet. Discuss the results. Discuss:
   - Were some perimeters harder to estimate?
   - Did your estimates get more accurate with practice?
   - What were some of the strategies you used in estimating?

   Prepare a new list, on chart paper or the bulletin board, entitled "What We Know About Perimeter." Write:
   ***Perimeter is the distance around something.***

3. Have students imagine they are inchworms crawling around the very edge of a tile. Have them run a finger around the sides of one tile. Ask: How far have you crawled? (*4 inches*) What is the perimeter? (*4 inches*) What is the area of the whole tile? (*1 square inch*) Have students look at the sketch and measurements on the Inchworm Investigations activity sheet.

4. Have students put 2 tiles side by side and find their perimeter and area. Ask: How far have you crawled? (*6 inches*) What is the perimeter? (*6 inches*) What is the area of the whole figure? (*2 square inches*) Students should draw their figure on the activity sheet and write their results.

5. Now have students make a figure with 3 tiles. Are all the figures the same? (*No. Some students will have a 1 by 3 rectangle and some will have a right-angle shape.*) Discuss what constitutes a different figure, and explain that the following figures are identical:

For both the rectangle and the right-angle shape ask: What is the perimeter? (*8 inches*) What is the area? (*3 square inches*) Have students add this information to their sheets. Ask: Will figures with the same area always have the same perimeter? (*Most students will think this is true.*)

6. Have students each use 4 tiles to complete the Inchworm Investigations activity. Discuss the results:
   - What methods did you use to find the perimeters?
   - Do you see any patterns?
   - Did figures with the same area always have the same perimeter?

   Add to the "What We Know About Perimeter" list.

   ***We can add the lengths of each side of a figure to get the perimeter.***

   ***Figures with the same area can have different perimeters.***

## Journal Writing

Draw a picture or diagram to explain the difference between area and perimeter.

## Extensions

- Have students investigate the figures that can be created with five tiles. (There are 12 possible configurations. Eleven have a perimeter of 12 units; one has a perimeter of 10 units.) These 5-tile figures are called *pentominos.*
- Have students make a large rectangle on graph paper, then draw a figure inside the rectangle with a perimeter larger than that of the rectangle. For example:

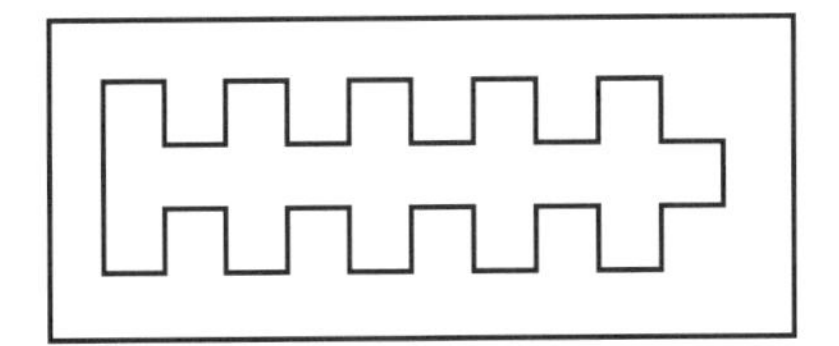

## Answers

### Inchworm Investigations

| | Perimeter | Area |
|---|---|---|
| 1 Tile | 4 in. | 1 sq. in. |
| 2 Tiles | 6 in. | 2 sq. in. |
| 3 Tiles | 8 in. | 3 sq. in. |
| | 8 in. | 3 sq. in. |
| 4 Tiles | 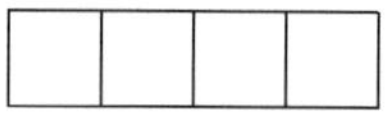 10 in. | 4 sq. in. |
| 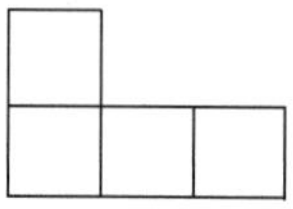 | 10 in. | 4 sq. in. |
| 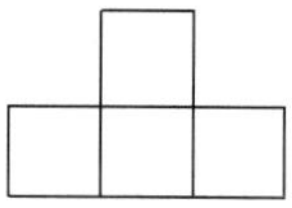 | 10 in. | 4 sq. in. |
| 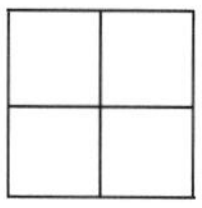 | 8 in. | 4 sq. in. |
| 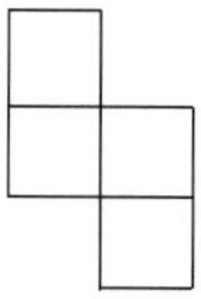 | 10 in. | 4 sq. in. |

Name Date

# STRING STRATEGIES

Estimate the perimeter, in centimeters, of an object in your classroom. Write the name of your object and your estimate in the chart.

Now use your string to measure the perimeter of the object. Then lay the string along your meter stick to find the perimeter in centimeters. Write the perimeter in the chart.

Now find 5 other objects. Estimate then measure their perimeters. Make sure the objects are not all rectangular.

| Object | Perimeter Estimate | Perimeter |
| --- | --- | --- |
| 1. | | |
| 2. | | |
| 3. | | |
| 4. | | |
| 5. | | |
| 6. | | |

Name Date

## INCHWORM INVESTIGATIONS

Imagine that an inchworm crawls around the outside edges of a shape made from square tiles. How far does the inchworm crawl? Make as many different shapes as you can with each number of tiles. Tiles cannot touch just at the corners:

*OK* *not OK*

Find the perimeter and then the area for each shape. Sketch each shape, and write your results in the chart. The first one is done for you.

| Sketch | Perimeter | Area |
|---|---|---|
| 1 Tile | 4 inches | 1 square inch |
| 2 Tiles | | |
| 3 Tiles | | |
| | | |
| 4 Tiles | | |
| | | |
| | | |
| | | |
| | | |

# Lesson 10: Rectangle and Square Shortcuts

**Developing a Formula for the Perimeter of a Rectangle**
**Developing a Formula for the Perimeter of a Square**

## Materials

meter sticks—1 per pair

cm rulers—1 per pair

1-cm graph paper (see p. 96)—2 sheets per student

geoboards and rubber bands—1 per student

Rectangle Shortcuts activity sheet—1 per student or pair

Square Shortcuts activity sheet—1 per student or pair

## Procedure

1. Review what students have learned about perimeter so far. Have pairs use meter sticks and centimeter rulers to complete the Rectangle Shortcuts activity. Each student should complete a separate sheet.

2. Discuss results:
   - Did you notice any patterns?
   - What shortcuts did you find for finding the perimeter of a rectangle?

   Review that we can write shortcuts in math language called *formulas.* A formula uses letters and symbols to explain relationships.

3. Have students, using $P$ for *perimeter,* $L$ for *length,* and $W$ for *width,* try to write on their activity sheets a formula for finding the perimeter of a rectangle. Have them start with $P =$ .

4. Discuss the formulas students create. There are two formulas commonly used: $P = 2L + 2W$ and $P = 2 \times (L + W)$. Add to the "What We Know About Perimeter" list:
   ***We can find the perimeter of a rectangle by using the formulas $P = 2L + 2W$ and $P = 2 \times (L + W)$.***

5. Have students make a chart on graph paper with the following headings:

   **Rectangle Sketch** **Length** **Width** **P =**

   Have them finish the formula heading with either $P = 2L + 2W$ or $P = 2 \times (L + W)$. Students create 6 different nonsquare rectangles, record the length and width in meters, and use the formula to find the perimeter. They use the sketches to check their accuracy.

6. Ask students: Do the same formulas work for squares? Have students use geoboards to complete the Square Shortcuts activity. Discuss results and students' formulas. Explain that since the sides of a square are of equal length, we don't have to use Length and Width. Instead, we use $s$ for *side.* The formula for the perimeter of a square is $P = 4s$. Add to the "What We Know About Perimeter" list:

   ***We can find the perimeter of a square by using the formula P = 4s.***

7. Have students make a chart on a new sheet of graph paper:

   **Square Sketch** **Side** **$P = 4s$**

   Students create 6 different squares on the graph paper, write the side lengths in centimeters, and use the formula to find the perimeter. They use the sketches to check their accuracy.

## Journal Writing

A recipe is a kind of formula. Write a formula for trail mix. Use $M$ for mix, $R$ for raisins, and $P$ for peanuts. Write an explanation of any other ingredients you choose to use. You might choose to have more of some ingredients than others in your mix.

## Extensions

- Have students bring ingredients and make their trail mix according to their formulas.
- Have students estimate, then measure, the perimeter of the school in paces. Repeat the process for the perimeter of the playground. Ask: Was one harder to estimate than the other? Why?
- Were all your measurements the same? How accurate do you think your results are? Why?

Name Date

# RECTANGLE SHORTCUTS

Find 6 rectangular objects in the classroom to measure with your meter stick or ruler. Write the measurements in the chart, and then find the perimeter.

| Object | Length 1 | Length 2 | Width 1 | Width 2 | Perimeter |
|---|---|---|---|---|---|
| 1. | | | | | |
| 2. | | | | | |
| 3. | | | | | |
| 4. | | | | | |
| 5. | | | | | |
| 6. | | | | | |

You know that you can add all the sides to find the perimeter of a rectangle. Can you find an easier way to find the perimeter of a rectangle—a shortcut? Write your shortcut here, and explain how you found it.

Name Date

## SQUARE SHORTCUTS

On your geoboard, make 4 different squares. All sides must be parallel to a geoboard edge. Draw each square you made, fill in the data, and find the perimeter. Look for a shortcut to the formula.

| Object | Length 1 | Length 2 | Width 1 | Width 2 | $P = 2L + 2W$ |
|---|---|---|---|---|---|
| 1. | | | | | |
| 2. | | | | | |
| 3. | | | | | |
| 4. | | | | | |

Does the formula for the perimeter of a rectangle work for squares? ________

Write a better formula for the perimeter of a square, and explain how you found it.

# Lesson 11: Quick Quilts

## Using Area and Perimeter Knowledge for Art and Sewing

### Materials

pictures of quilts (books are good sources)

$\frac{1}{4}$-in. graph paper (see p. 93)—several sheets per student

markers or crayons

scissors

*optional:* fabric and fabric crayons

### Procedure

1. Show pictures of various types of quilts. Students may have samples at home to bring in and share. Discuss various kinds of quilts such as *appliqué,* in which a design is sewn onto a quilt block, and *patchwork,* in which scraps of fabric are sewn together to create the block.

2. Discuss making a class quilt. Each student could make a patchwork quilt block on graph paper, from construction paper, or from fabric using fabric crayons. If you sew or have willing volunteers, considering creating a fabric quilt; they make wonderful donations to homeless or women's shelters.

3. With the class, decide on the finished quilt size. This will probably be determined by where the quilt will be displayed or how it will be used (if it is made of fabric). Next decide on the number of blocks. This will depend on the number of students in the class, although extra blocks could be added if necessary. Have students draw possible configurations of the quilt—for example, 1 block by 24 blocks—on paper. Have them decide on the best configuration.

   Discuss what size block would be best. Have students figure the finished quilt size for several sizes of block. Example: For a 4 block by 6 block quilt, a block size of 8" by 8" makes a 32" by 48" quilt. Note: In planning for a fabric quilt, remember to add seam allowances before cutting the fabric.

4. Plan for a border or fabric binding. Ask: What are the perimeters of the various arrangements of blocks we've suggested? What arrangements give the largest perimeters? The smallest?

5. If you're constructing a fabric quilt, have students research the cost of binding and figure the amount they will need. Binding is usually in packages of 3 yards, so this is practice in division with remainders. Graph the number of packages of binding needed for each of the different quilt perimeters. A bar graph would effectively show this.

6. Explore the area of the border. Have students sketch a quilt and border. Example: A 32" by 48" quilt with a 3" border will need two 38" by 3" strips and two 48" by 3" strips of border. Ask: What is the area of the border? (*228 + 288 = 516 square inches*) How many sheets of 9" by 12" construction paper will we need for the border? (*4.8, which means using 5*)

7. There are many ways to arrange individual blocks within the quilt. Have students explore how many ways they can arrange four tiles of different colors into a square. Have them estimate the number. Labeling the squares and making an organized list helps (e.g. 1234, 1243, 1324, 1342, 1423, 1432, . . . ). For example:

| 1 | 2 |
|---|---|
| 3 | 4 |

For four squares, the possible number is $4 \times 3 \times 2 \times 1 = 24$, or 4!, called *four factorial.* If there are 10 students in the class, there are 3,628,800 ways of arranging the blocks. This would be quite a series of configurations for the class to try!

8. Students can now use graph paper to make rough sketches of their designs for their blocks. Students can cut out their graph paper blocks, and the class can decide on a whole-quilt arrangement before they make their final blocks. Some students may find they want to change the color or design of their blocks once they see how they fit into the overall quilt.

## Journal Writing

Think of a piece of clothing fabric from your past that has meaning for you and that might be used in a patchwork quilt. The fabric does not have to be something of yours. Describe it and write a story of your memory.

## Extensions

- Have students measure the area of a large bulletin board. Ask: If you want to completely fill the bulletin board with rectangular quilt blocks, what dimensions could your blocks be?
- Have each student use graph paper to design a symmetrical patchwork pattern, then design a whole quilt of these squares. Students must choose a scale that will allow them to see the overlying design before they create the blocks. You could tie this in with a unit on tessellation.
- Read *The Patchwork Quilt,* by Valerie Flournoy (New York: Dial Books for Young Readers, 1985), to the class. Tanya, her mother, and her grandmother share family stories and

memories as they work on a quilt made from clothing from family members. Have each student draw a meaningful symbol for himself or herself on graph paper or fabric quilt blocks. Use large blocks and put them together for quilts, or use small blocks and use them for illustrations in student-written books. Have students write a chapter about each square.

- For more on factorials, read *Anno's Mysterious Multiplying Jar* (Masaichiro Anno and Misumasa Anno, illustrated by Misumasa Anno; New York: Philomel Books, 1983) to the class. This is a short book and beautifully illustrated.

# Lesson 12: Pinwheels and Propellers

## Relating Area and Perimeter of Rectangles—Holding Area Constant

## Materials

calculators

$\frac{1}{4}$-in. graph paper (see p. 93)—1 sheet per pair

scissors

tiles—16 per student

Pinwheels and Propellers activity sheet, pages 1 to 3—1 set per student

## Procedure

1. Discuss students' ideas of the relationship between area and perimeter:
   - Can figures with the same area have different perimeters?
   - Will figures with the same perimeter have the same area?

   List predicted results on the board.

2. Discuss whirligigs, three-dimensional figures on sticks, such as pinwheels and airplane models with rotating propellers, that have a part that moves in the wind. Many people use whirligigs as lawn or garden ornaments. Have students complete the Pinwheels and Propellers activity.

3. Discuss students' discoveries and conclusions. Add to the "What We Know About Area" list:
   ***For rectangles with the same area, the larger the difference between the lengths of the sides, the larger the perimeter.***
   ***A square has the smallest perimeter for any given rectangle.***

4. Have students predict the smallest and largest perimeters possible for rectangles made from 9 tiles. Have them test their predictions by making the rectangles. Repeat for rectangles made from 16 tiles. Ask: What do you notice about the area of rectangles with the smallest possible perimeter? (*They are square number areas.*)

5. Have students cut twelve 1" by 1" squares from graph paper. Ask: What's the largest perimeter you can find for a rectangle with an area of 12 square units? Students should find they can cut the squares in halves or fourths and reassemble them into new rectangles with perimeters of 49" or 96.5".

## Journal Writing

Write a conversation between the area and the perimeter of a rectangle.

Explain what is most important to understand about perimeter and area of rectangles.

## Extensions

- Have students graph their perimeter results for all rectangles with an area of 12 square units. The $x$-axis should be the length and the $y$-axis the width of the rectangle. Students should plot both positions for each rectangle; for example, 1 by 12 and 12 by 1. The resulting curve is a hyperbola.
- If fencing costs $3 per foot, how much will students pay to fence their rectangle choices? What happens if the fencing only comes in 6-foot lengths?

## Answers

### Pinwheels and Propellers

Area = 12 square feet

| Width | Length | Perimeter | Area |
|---|---|---|---|
| 1 ft. | 12 ft. | 26 ft. | 12 sq. ft. |
| 2 ft. | 6 ft. | 18 ft. | 12 sq. ft. |
| 3 ft. | 4 ft. | 14 ft. | 12 sq. ft. |

Area = 36 square feet

| Width | Length | Perimeter | Area |
|---|---|---|---|
| 1 ft. | 36 ft. | 74 ft. | 36 sq. ft. |
| 2 ft. | 18 ft. | 40 ft. | 36 sq. ft. |
| 3 ft. | 12 ft. | 30 ft. | 36 sq. ft. |
| 4 ft. | 9 ft. | 26 ft. | 36 sq. ft. |
| 6 ft. | 6 ft. | 24 ft. | 36 sq. ft. |

Name Date

## PINWHEELS AND PROPELLERS, PAGE 1

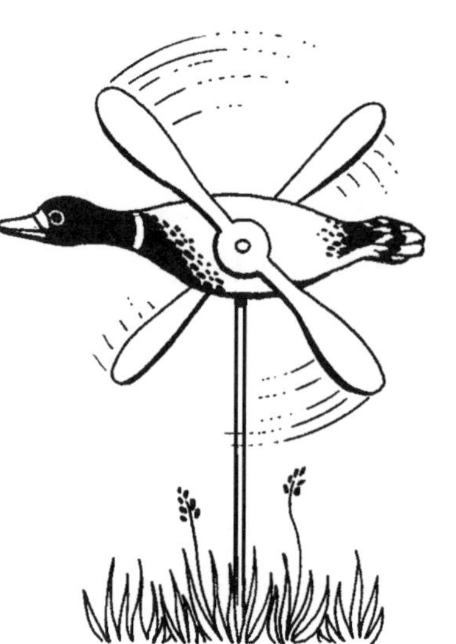

Ken and Reiko want to display their collection of garden whirligigs. They'll need 1 square foot of space for each whirligig so the wind can move them freely. They want to build a fence around their collection. Fencing is expensive, but a long, thin line of whirligigs can be seen better than a block of whirligigs. Ken and Reiko decide to see what shapes are possible.

Find the rectangular shapes they can use for 12 whirligigs. Sketch and label each shape, and fill in the chart. The first one is done for you. Try to see if there are any patterns.

For 12 whirligigs, the area = 12 square feet.

| Sketch | Width | Length | Perimeter | Area |
|---|---|---|---|---|
| 1. • • • • • • • • • • • • | 1 ft. | 12 ft. | 26 ft. | 12 sq. ft. |
| 2. | 2 ft. | | | |
| 3. | 3 ft. | | | |

Name Date

# PINWHEELS AND PROPELLERS, PAGE 2

Now find the rectangular shapes Ken and Reiko can use for 36 whirligigs.

For 36 whirligigs, the area = 36 square feet.

| Sketch | Width | Length | Perimeter | Area |
|---|---|---|---|---|
| 1. | 1 ft. | | | |
| 2. | 2 ft. | | | |
| 3. | 3 ft. | | | |
| 4. | 4 ft. | | | |
| 5. | 6 ft. | | | |

Name Date

## PINWHEELS AND PROPELLERS, PAGE 3

What patterns did you find?

What do you notice about the length and width of rectangles that have large perimeters for their area?

If two rectangles have the same area, what shape of rectangle has the larger perimeter for that area?

What do you notice about the length and width of rectangles that have small perimeters for their area?

If two rectangles have the same area, what shape of rectangle has the smallest perimeter for that area?

What rectangular area would you choose to display 12 whirligigs?

Why?

# LESSON 13: MAXIMICING A CAKE

### Relating Surface Area and Perimeter of Rectangular Solids

## Materials

calculators

cubes, preferably interlocking—16 per pair or group

Maximicing a Cake activity sheet, pages 1 to 4—1 set per pair

*optional:* the Designer's Dilemmas activity sheet could be used after the cake activities

## Procedure

1. Have students imagine that a cube is a 3 inch by 3 inch by 3 inch piece of cake, and that they want to ice the piece before eating it. Show students how to sketch a network to represent the cake and iced surfaces.

|   | S |   |
|---|---|---|
| S | T | S |
|   | S |   |

S = side

T = top

2. Have students enter data on page 2 of the Maximicing a Cake activity sheet as they discuss:
   - How many surfaces will get iced? (*5: 1 top and 4 sides*)
   - What is the area of one side? (*9 sq. in.*)
   - How much area in all can be iced? (*5 surfaces × 9 sq. in. = 45 sq. in.*)

   The smallest amount of icing they can make is $\frac{1}{2}$ cup, which covers 90 square inches of cake. The icing recipe can be increased $\frac{1}{2}$ cup at a time. Discuss:
   - How much icing will you need? ($\frac{1}{2}$ *cup. There will be some left over.*)
   - What is the perimeter of the cake? (*12 in.*)

3. Remind students that cakes are first iced, then cut into pieces. Have students predict how many surfaces will get iced if they use two cubes for the cake. Then have them put two cubes together to make a rectangular prism. Students use a network to sketch the cake and enter the data on page 2 as you discuss:
   - How many surfaces will get iced? (*8: 2 tops and 6 sides*)

There were 5 surfaces to frost with one cube. There are now twice as many cubes.

- Why aren't there twice as many surfaces to be frosted? (*Two of the sides are together and don't get iced.*)
- How much area in there to ice? (*8 surfaces* $\times$ *9 sq. in. = 72 sq. in.*)
- How much icing is needed? ($\frac{1}{2}$ *cup. There will be some left over.*)
- What is the perimeter of the cake? (*18 in.*)

4. Discuss a 4-cube cake. Ask: What size cakes are possible with 4 cubes? (*2* $\times$ *2, 1* $\times$ *4*) Have students work through the 4-cube activity, sketching the cake and charting their results on page 2.

5. Go over page 1 with the class. Make sure students know the conditions. Have students complete pages 3 and 4. Discuss their findings.

## Journal Writing

Write a letter to an adult explaining why you want a long, narrow cake. Be convincing.

## Extensions

- Have students do the Designer's Dilemmas activity.
- Have students figure out the largest cake they could have on top of their desks. Each piece still remains 3" by 3"; lengths and widths must be in multiples of 3. How much icing will the cake need?

## Answers

### Page 1

| Amount of Icing | Area of Cake It Covers |
| --- | --- |
| $\frac{1}{2}$ cup | 90 sq. in. |
| 1 cup | 180 sq. in. |
| $1\frac{1}{2}$ cups | 270 sq. in. |
| 2 cups | 360 sq. in. |
| $2\frac{1}{2}$ cups | 450 sq. in. |

**Page 2**

| Shape of Cake | Number of Tops | Area of Tops | Number of Sides | Area of Sides | Area | Total Perimeter | Icing Needed |
|---|---|---|---|---|---|---|---|
| $1 \times 1$ | 1 | 9 sq. in. | 4 | 36 sq. in. | 45 sq. in. | 12 in. | $\frac{1}{2}$ cup |
| $1 \times 2$ | 2 | 18 sq. in. | 6 | 54 sq. in. | 72 sq. in. | 18 in. | $\frac{1}{2}$ cup |
| $2 \times 2$ | 4 | 36 sq. in. | 8 | 72 sq. in. | 108 sq. in. | 24 in. | 1 cup |
| $1 \times 4$ | 4 | 36 sq. in. | 10 | 90 sq. in. | 126 sq. in. | 30 in. | 1 cup |

**Page 3**

| Shape of Cake | Number of Tops | Area of Tops | Number of Sides | Area of Sides | Area | Total Perimeter | Icing Needed |
|---|---|---|---|---|---|---|---|
| $4 \times 4$ | 16 | 144 sq. in. | 16 | 144 sq. in. | 288 sq. in. | 48 in. | 2 cups |
| $2 \times 8$ | 16 | 144 sq. in. | 20 | 180 sq. in. | 324 sq. in. | 60 in. | 2 cups |
| $1 \times 16$ | 16 | 144 sq. in. | 34 | 306 sq. in. | 450 sq. in. | 102 in. | $2\frac{1}{2}$ cups |

**Page 4**

1. Each cube has a top, so the number of tops equals the number of cubes.
2. When you put two cubes together you can't ice the touching sides so you can't count these sides. The 4-cube and 16-cube cakes have more than one possible shape.
3. The square shape has the smallest perimeter; the larger the difference between the length and the width, the larger the perimeter.
4. You need to know the total area to be covered and that $\frac{1}{2}$ cup of icing covers 90 square inches.
5. For a given number of pieces, the longest cake has the most icing because there are more sides available to be iced.
6. The cake with the 1 by 16 shape wastes no frosting. This cake's surface area is 450 square inches, exactly $2\frac{1}{2}$ cups of icing.
7. A 5 by 5 cake would have the least number of sides facing out (because it would have the smallest perimeter) and would require the least amount of icing.

Name Date

# MAXIMICING A CAKE, PAGE 1

Voshan is having a birthday party. He wants to invite 15 guests and will need _____ pieces of his favorite carrot cake. He's *very* fond of the icing and wants to make sure his guests get as much icing on their cake as possible.

His mother is a baker and has rectangular baking pans in many sizes and dimensions. She's promised to let Voshan design the shape of the cake as long as he understands some conditions:

- Each of the 16 pieces will be 3 inches by 3 inches by 3 inches The cake will be rectangular.
- His mother will make the cake in one layer, take the cake from the pan, and ice the sides and top.
- After the cake is iced, she will cut it into pieces.
- Voshan will need to tell her how much icing is needed. Her recipe makes $\frac{1}{2}$ cup of icing, which will cover about 90 square inches of cake. She can increase the recipe $\frac{1}{2}$ cup at a time to make more.

| Amount of Icing | Area of Cake It Covers |
|---|---|
| $\frac{1}{2}$ cup | 90 sq. in. |
| 1 cup | |
| $1\frac{1}{2}$ cups | |
| 2 cups | |
| $2\frac{1}{2}$ cups | 450 sq. in. |

1. Fill in the Icing chart above.
2. Use 16 cubes to experiment with cake shapes.
3. Draw your cakes, and enter your data in the chart on page 3.
4. Look for patterns. Find which shape cake will give Voshan and his guests the most icing.
5. Answer the questions on page 4.

Name　　　　　　　　　　　　　　　　　　　　Date

# MAXIMICING A CAKE, PAGE 2

| Shape of Cake | Number of Tops | Area of Tops | Number of Sides | Area of Sides | Total Area | Perimeter | Icing Needed |
|---|---|---|---|---|---|---|---|
| 1-Cube Cake<br>$1 \times 1$ | | | | | | | |
| 2-Cube Cake<br>$1 \times 2$ | | | | | | | |
| 4-Cube Cake<br>$2 \times 2$ | | | | | | | |
| $1 \times 4$ | | | | | | | |

Name Date

# MAXIMICING A CAKE, PAGE 3

| Shape of Cake | Number of Tops | Area of Tops | Number of Sides | Area of Sides | Total Area | Perimeter | Icing Needed |
|---|---|---|---|---|---|---|---|
| 16-Cube Cake | | | | | | | |
| $4 \times 4$ | | | | | | | |
| $2 \times 18$ | | | | | | | |
| $1 \times 16$ | | | | | | | |

Name Date

# MAXIMICING A CAKE, PAGE 4

1. What do you notice about the number of cubes and the number of tops for each cake?

2. Why are there different side areas for the 4-cube cakes and the 16-cube cakes?

3. For the 4-cube and 16-cube cakes, what do you notice about the perimeter and the shape of the cakes?

4. What do you need to know to find the amount of icing needed to ice a particular cake?

5. What shape of cake has the most icing? Why?

6. Which cake has the least amount of icing left over? How do you know?

7. If you had a 25-piece cake, which shape do you predict would have the least icing? Why?

Name Date

## DESIGNER'S DILEMMAS

The architects that design buildings have to consider many things. Think about the cake activities you've just done as you answer these questions.

1. What building shape would be allow the most possible light to come through the windows?

2. What building shape would give the most economical heating and cooling?

3. What building shape require the least amount of paint to paint the outside?

4. What building shape would require the least amount of carpeting to carpet the floor inside?

Imagine that you are an architect. Design a building that will be energy efficient as well as letting in as much light as possible through the windows. You are not limited to a rectangular building this time. Draw your building below, then explain your reasoning on the back of this sheet.

# Lesson 14: Native American Designs

**Exploring Triangular Area**

**Finding Area with Half Units**

## Materials

geoboards and rubber bands, or geoboard paper (see p. 92)—1 per student

1-cm graph paper (see p. 96)—1 sheet per student

colored markers

transparency of Navajo and Seminole Designs (see p. 98)

Design Units activity sheet—1 per student

*optional:* pictures of the designs of Navajo, Seminole, or other cultural groups that incorporate rectangles and triangles

## Procedure

1. Explain that many cultures have used triangles and rectangles regularly in design. Project the transparency of Navajo and Seminole Designs. The first example is part of a design that might appear on a woven Navajo rug. Ask: How would knowing the area of each color be useful to the weaver? The second design shows a pattern on a band of fabric that might decorate a Seminole Indian skirt. Ask: How would knowing the area of each color be useful to the sewer?

Navajo-Type Design

Seminole-Type Design

2. Have students make a 1 by 1 square on their geoboards. This will be one unit. Now have them use another rubber band to divide the square into two triangles. Ask: Are these triangles congruent? Discuss that each triangle has one right angle, each height is one unit, and each base is one unit; they are congruent.

3. Have students make figures with areas of $1\frac{1}{2}$ square units. Ask: How many different designs are there? Repeat with areas of 2 and $3\frac{1}{2}$ square units. Ask: Which area has the greatest number of designs? Why?

4. Have students complete the Design Units activity sheet.

5. Have students share their results and designs with a partner. Ask what they discovered about area of triangles and rectangles. Discuss and add to the "What We Know About Area" list:
   ***Two triangles can be put together to make a rectangle.***
   ***Each triangle is half the area of its rectangle.***
   ***Fractional units can be added to get total area.***

6. Have students make a design 5 centimeters wide and 15 centimeters long on graph paper, repeating design elements of rectangles and triangles. How much area will each design cover? Display the results.

## Journal Writing

You have a long, rectangular piece of cloth for a belt, and you want to draw 20 evenly spaced triangles on the belt. How will you do it?

## Extension

- Have students research the designs of a number of Native American cultures. Then have them make and color designs in the style of one of them. Have them figure the area of each color in their design.

## Answers

A: 2 sq. units; B: $2\frac{1}{2}$ sq. units; C: 5 sq. units; D: 7 sq. units; E: $4\frac{1}{2}$ sq. units; F: 7 sq. units.

Name Date

# DESIGN UNITS

Find the areas of designs A, B, and C.

Area = ______ sq. units Area = ______ sq. units Area = ______ sq. units

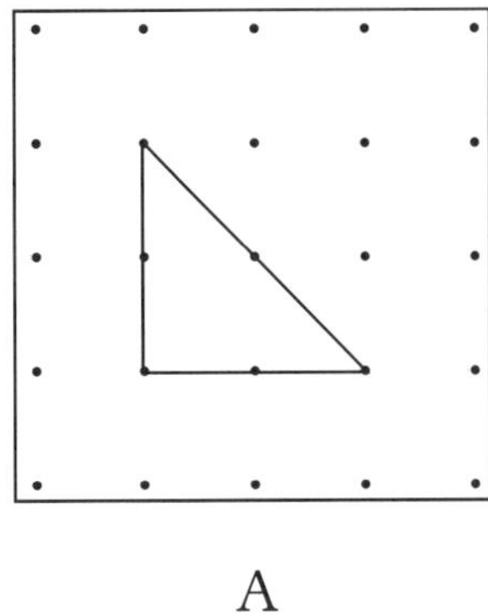

A

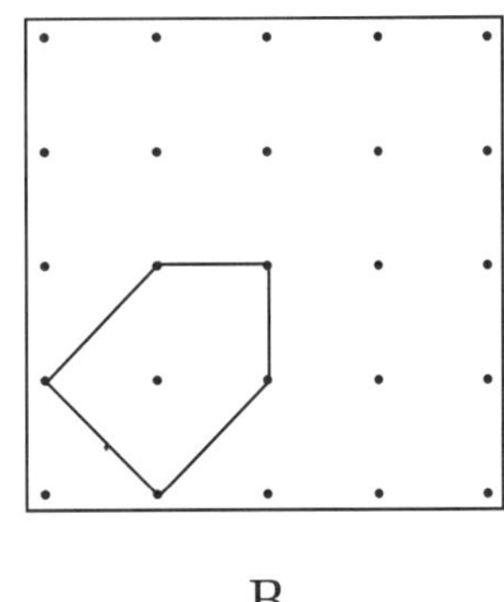

B

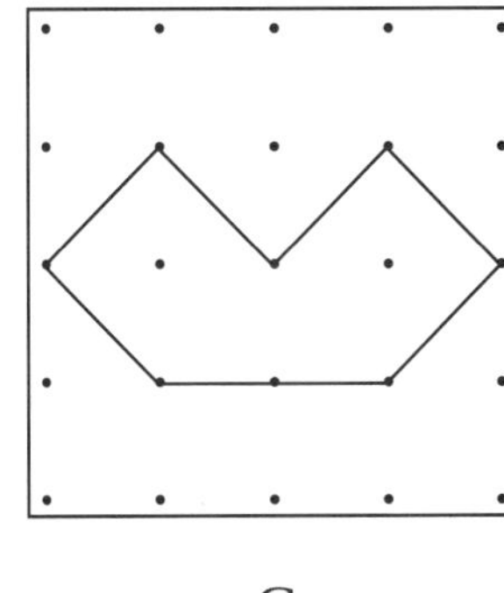

C

Estimate the areas of designs D, E, and F. List your estimates from largest to smallest: 

Now find the actual areas. Which area was hardest to estimate? ______________

Area = ______ sq. units Area = ______ sq. units Area = ______ sq. units

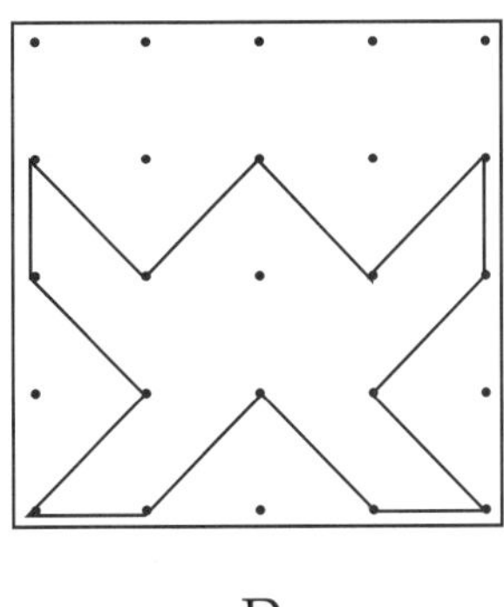

D

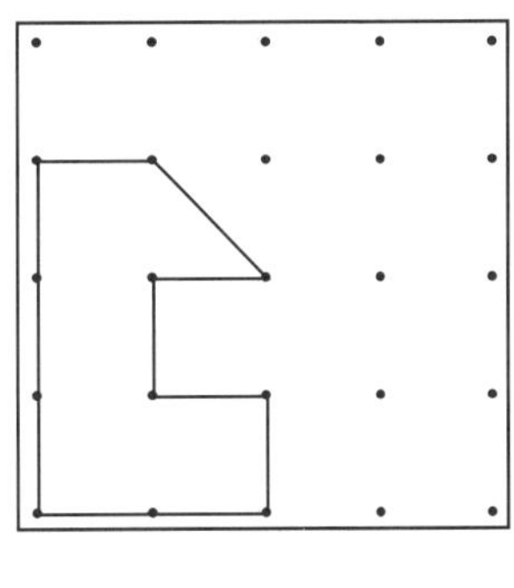

E

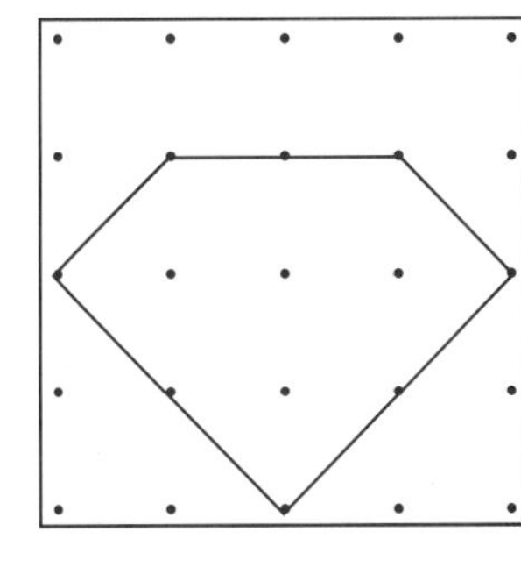

F

Draw three different designs with an area of $5\frac{1}{2}$ square units.

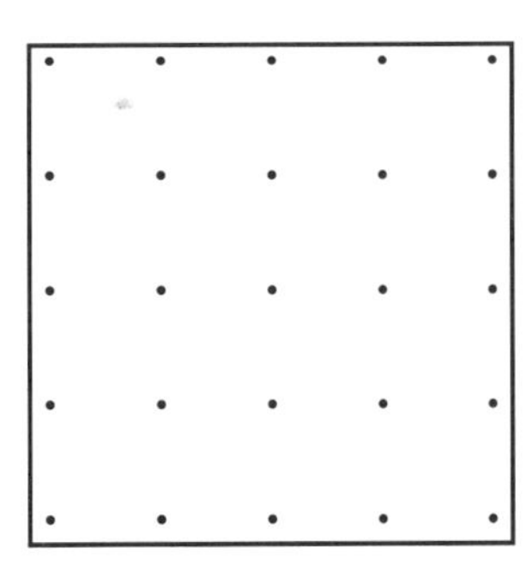

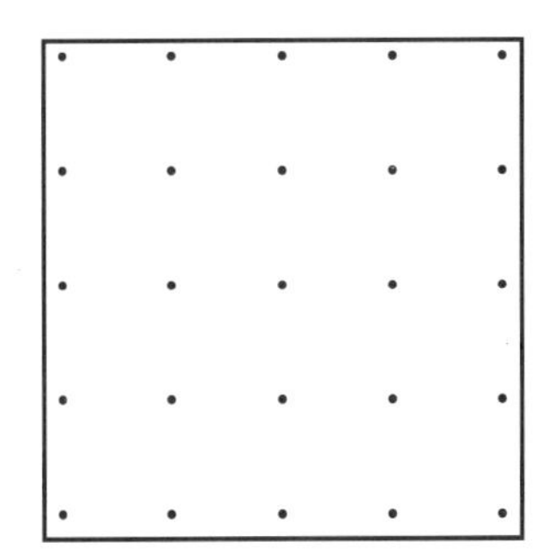

# Lesson 15: Palmistry

## Estimating Area Within a Closed Curve

### Materials

$\frac{1}{4}$-in. or 1-cm graph paper (see pp. 93 and 96)—1 sheet per student

transparency of the graph paper used—1 per student

scissors

writing paper—1 sheet per student

plain paper—1 sheet per student

### Procedure

1. Have students compare the amount of skin on their palms with that on the rest of their bodies. What fraction or percentage of their total body skin do they think their palms contain? Have them use graph paper and trace around one of their hands, stopping at the wrist. Fingers and thumb should be touching.

2. Direct students to find the area contained in this closed curve and to explain their method and findings in writing.

3. Discuss methods students used. What did they decide to do about squares only partially contained within the boundary? Many will have used fractional parts of squares.

4. Now have students place a dot in the center of each square. If the dot is within the boundary, they count the square. If the dot is outside the boundary, they ignore the square. If the boundary is on the dot, they count half a square. Discuss:
   - Was this method of calculating area easier?
   - Is it more accurate, or not? Why?

5. Students could also try the method of cutting the region into squares, or rectangles that contain only whole squares. Areas with partial squares are then reassembled into whole squares.

   Add to the "What We Know About Area" list:

   ***We can estimate the area within a closed curve by counting squares and fractions of squares.***

6. How could students get a more accurate estimate? Have students trace one of their hands on plain paper, place the transparency grid over the tracing, and figure the area. Next have them move the transparency, figure the area again, and take an average of their two results. Ask:
   - How does this method compare to the other methods we used?
   - Do you think it's more accurate? Why?

   Discuss: Another way to get a more accurate estimate is to use graph paper with smaller divisions.

7. The skin on both palms is about 1% of the body's skin. Surgeons in burn centers use this knowledge to estimate the amount of a patient's body that is burned and to plan for skin grafting. Have students use the area of their palm skin to estimate the area of all their skin. Ask: How might the estimate of your total-skin area be affected if the palm-skin area estimate is inaccurate? Have students use a slightly larger or smaller palm skin area to re-estimate their total skin area.

## Journal Writing

Describe something about area that you now know or understand better. What helped you understand it?

## Extensions

- Do students with larger hands have larger feet? Have students find the area of their feet and plot their results on a whole-class graph. Example: List students' names along the bottom of the graph, starting with the student with the smallest hand. Students use one color marker to make a bar showing their hand size, and another color marker to make a bar showing their foot size. Ask: What can you conclude from these results?
- Do students' hand sizes correlate with their heights? Have them plot class results as a scattergram. Is this different for boys than for girls? Are results different for different ages?

# Lesson 16: Soap and Water

### Estimating Area Within a Closed Curve

## Materials

calculators

clear plastic wrap, enough to cover the graph paper

small containers—3 per group

1 tablespoon dishwashing detergent

food coloring, in 3 colors

$\frac{1}{4}$-in. graph paper (see p. 93)—1 per group

masking tape

3 measuring cups, at least 1 cup each

$\frac{1}{2}$-teaspoon measuring spoons—1 per group

rulers—1 per group

paper towels

water at room temperature

Soap and Water activity sheet—1 per group

*optional:* cooking oil

## Procedure

*Ahead of time:* Have students put 1 cup of room-temperature water in each of the three measuring cups and add a different color of food coloring to each. They add $\frac{1}{8}$ teaspoon of dishwashing detergent to one cup and stir until it is mixed. They add $\frac{1}{2}$ teaspoon of detergent to another cup and stir until it is mixed. There should not be any bubbles.

1. Have students estimate the area, in square inches, that $\frac{1}{2}$ teaspoon of water will cover. Write the estimates on the board. Discuss different ways to measure this. Explain that students will be measuring this by dropping water onto a grid. Discuss variables they might want to control (e.g. the temperature of the water, source of the water, the height from which the water is dropped).

2. Have students place a sheet of graph paper on a flat surface, cover it with plastic wrap, and tape the edges to make a wrinkle-free surface. They then carefully drop $\frac{1}{2}$ teaspoon of water onto the plastic wrap from a height of 2 inches, estimate the area covered, and write under Observations anything they notice about the water. Suggest they repeat this twice more,

drying the plastic wrap carefully between experiments. They chart their data and observations on the Soap and Water activity sheet and use calculators to find the average area. Finally, they add their average area to a class chart.

3. Discuss findings:
   - Are all the results the same?
   - How could you decide which result is probably correct?

   Take an average of all the areas, and compare with the predictions.

4. Students may have noticed that the surface of the water was not flat. Explain that liquids are made up of small particles called *molecules.* Molecules are too small to be seen except with very powerful microscopes. The molecules at the surface of a liquid are attached to each other. If you were to pull on the ends of a rope, you would be putting *tension* on the rope. The layer of molecules at the surface of water is also under a kind of tension, called *surface tension.* Surface tension is what holds the water in a drop shape.

5. Ask students to predict what will happen if soap is added to the water. How much area will $\frac{1}{2}$ teaspoon of water cover then? Write predictions on the board. Have students repeat the activity using the water to which $\frac{1}{8}$ teaspoon of detergent has been added.

6. Discuss results and observations. The detergent lowers the surface tension of the water, making it "wetter." The water is flatter if viewed from the side, and the drop spreads out to cover a larger area.

7. Have the class predict what will happen when four times as much as much detergent as before is added to a cup of water. Have class repeat the activity using the water to which $\frac{1}{2}$ teaspoon of detergent was added. Ask:
   - How accurate were your predictions this time?
   - What have you learned about surface tension and area?

## Journal Writing

Design another experiment to investigate surface tension and area. Start with "What if . . ."

## Extensions

- Have several students create bar graphs with their own data, while other students make a bar graph using the whole-class averages. How do the students' individual graphs compare with the whole-class graph?

- Have students put one or two drops of cooking oil on their plastic wrap and a drop of colored water next to the oil so that the drops touch. What do students observe? (*The two liquids do not mix.*) On the other side of the oil, have students put a drop of water to which detergent has been added. This drop should be a different color from the first. The second drop will mix with the oil, the oil's surface tension will be lowered, and the first drop will also mix with the oil. Does this have implications for proper hand washing?
- Use rainwater, or water that has been softened, for this activity, and compare findings with previous results.

Name Date

# SOAP AND WATER

For each sample, hold your spoon 2 inches above the surface and drop a $\frac{1}{2}$ teaspoon of water onto your plastic wrap. Estimate the area covered by the water. Record your estimate and observations in the chart. Wipe your plastic wrap dry. Repeat the procedure two more times, then find the total area covered in all 3 trials. Divide the total by 3 to get the average area covered.

**Sample 1: No Detergent**

| | Area | Observations |
|---|---|---|
| Trial 1 | | |
| Trial 2 | | |
| Trial 3 | | |
| Total | | |
| Average | | |

**Sample 2: $\frac{1}{8}$ Teaspoon of Detergent per Cup of Water**

| | Area | Observations |
|---|---|---|
| Trial 1 | | |
| Trial 2 | | |
| Trial 3 | | |
| Total | | |
| Average | | |

**Sample 3: $\frac{1}{2}$ Teaspoon of Detergent per Cup of Water**

| | Area | Observations |
|---|---|---|
| Trial 1 | | |
| Trial 2 | | |
| Trial 3 | | |
| Total | | |
| Average | | |

# Lesson 17: Houses for Sale

**Developing a Formula for the Area of a Right Triangle**

## Materials

calculators

geoboards and rubber bands

geoboard paper (see p. 92)—1 sheet per student

scissors

Houses for Sale activity sheet, pages 1 to 2—1 set per student

Builder's Challenge activity sheet—1 per student

## Procedure

1. Propose the following scenario: A builder wants to divide a piece of land into lots for sale. She would like to build a house on each lot and sell the lot and house together. Since she can make $30,000 on each house, she wants to build as many houses as possible. However, the city code states that each lot with a house on it must cover an area of at least 10,000 square feet. A stream runs through some of the land. Since home owners will not want to cross the stream to get to the rest of their property, she plans to use the stream as a boundary line. The property owners will own the water rights halfway out into the stream.

2. Discuss the Houses for Sale activity sheets. Have students complete them.

3. Discuss students' findings. Compare methods students used to find area. Most should see that the two triangles are congruent; they can find the area of the original rectangle and take half of that for each triangle.

4. Remind students that *formulas* are math language shortcuts that describe patterns or relationships. Review the formula for the area of a rectangle: $A = L \times W$. Discuss that when working with triangles, we use the terms *base* (written $b$) instead of length, and *height* ($h$) instead of width. Have students suggest a formula for the area of a right triangle, remembering to substitute the word *base* for length and *height* for width. After discussing the proposed formulas, tell students that the formula for the area of a right triangle is commonly written $A = \frac{1}{2}bh$.

5. Have students use geoboards to make right triangles and find their areas. All bases must be parallel to a side of the geoboard. If necessary, help students see they can "box in" each triangle to make a rectangle, find the area of the rectangle, then find half of that. Students should draw each triangle on geoboard paper and label the base and height. Next to each triangle,

they write $A = \frac{1}{2}bh$ and find the area by substituting for $b$ and $h$. Discuss whether the geoboard and formula results were the same for each triangle.

6. Add to the "What We Know About Area" list:
   ***The two triangles formed by a rectangle's diagonal are congruent, and each triangle has half the area of the rectangle.***
   ***The formula for area of a right triangle is $A = \frac{1}{2}bh$.***

7. Have students complete the Builder's Challenge activity sheet. There are several different ways to divide each example to find the area. Caution students that any triangles used in these divisions must have bases and heights parallel to an edge of the geoboard pictured. The hypotenuse is not equal to either the base or the height; the diagonal of a 1 by 1 square does not equal 1. Students who finish early should be encouraged to find another way to find each area.

## Journal Writing

Explain why it is important to know how to find the area of a right triangle.

## Extensions

- How else could the builder divide her land to have the right-sized lots and make more money by building more houses?
- Challenge students to use what they know to find areas of isosceles triangles.

## Answers

### Houses for Sale

1. A = 20,000 sq. ft.; B = 22,500 sq. ft.; C = 15,000 sq. ft.

3–5. They are congruent triangles. They each have a right angle.
Each has half the area of their rectangle.

6. The length, the width, a right angle, and that each hypotenuse is the diagonal of its rectangle.
7. A1 = 10,000 sq. ft.; A2 = 10,000 sq. ft.; B1 = 11,250 sq. ft.; B2 = 11,250 sq. ft.; C1 = 7,500 sq. ft.; C2 = 7,500 sq. ft.

9. Four

### Builder's Challenge

1. $7\frac{1}{2}$ sq. units
2. 11 sq. units
3. $6\frac{1}{2}$ sq. units
4. 6 sq. units
5. 8 sq. units
6. 9 sq. units
7. $10\frac{1}{2}$ sq. units
8. 12 sq. units
9. 9 sq. units

Name Date

## HOUSES FOR SALE, PAGE 1

The builder wants as many lots as possible. Each lot with a house must cover at least 10,000 square feet. She will use the stream as a boundary between lots.

1. The builder first divided the land into the rectangles shown on page 2. Find the area of each rectangle:

   A ________________ B ________________ C ________________

2. The stream divides each rectangle into two lots. Cut out each rectangle, then cut along the stream.

3. Write everything you notice about shapes A1 and A2.

4. Write everything you notice about shapes B1 and B2.

5. Write everything you notice about shapes C1 and C2.

6. Name three things all the triangles share with their original rectangle.

7. Find the areas: A1 ____________ A2 ____________ B1 ____________

   B2 ____________ C1 ____________ C2 ____________

8. Describe how you found the areas.

9. How many lots can the builder put houses on if she keeps the land divided this way? (Don't forget that there is a minimum number of square feet for a lot with a house.)

Name Date

# HOUSES FOR SALE, PAGE 2

This illustration shows the piece of land that the builder wants to divide into lots, and the stream that runs through the piece of land.

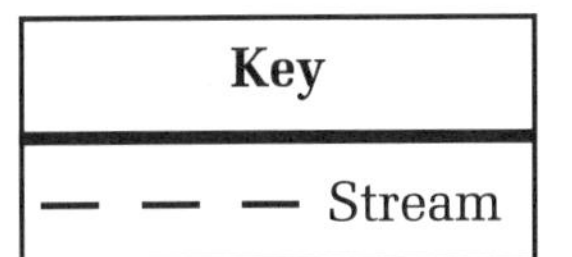


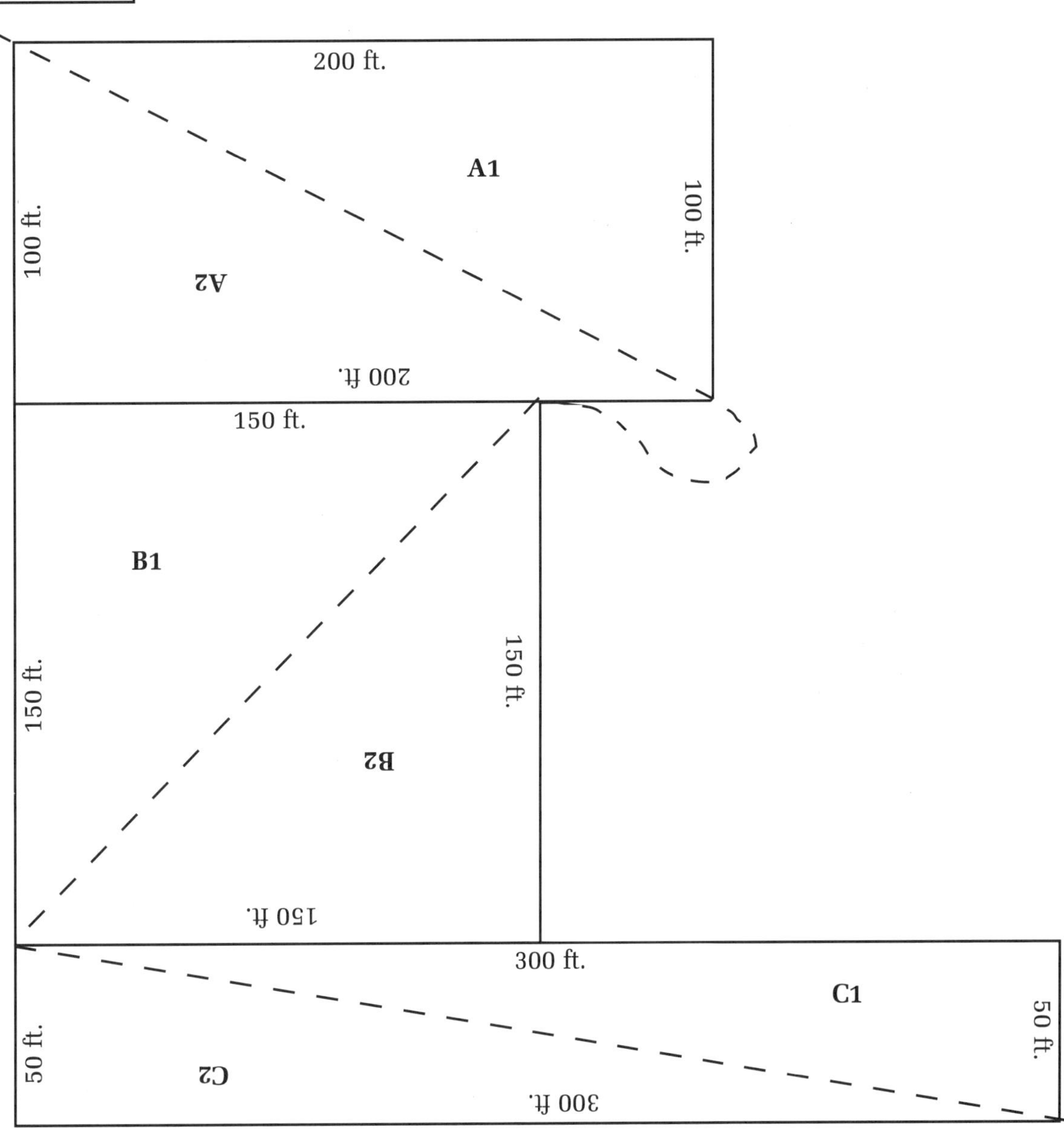

Name Date

## BUILDER'S CHALLENGE

Find the area of each piece of land. Label the area in square units. If you divide the land into smaller shapes, show your divisions, and show the area of each smaller pieces as well as the total area.

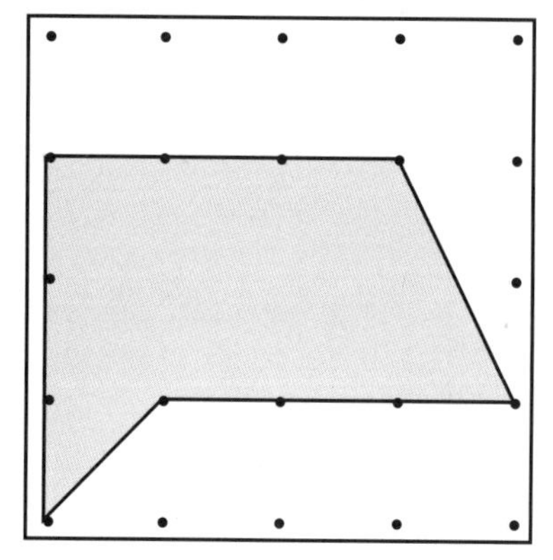

1. $A =$ ____________

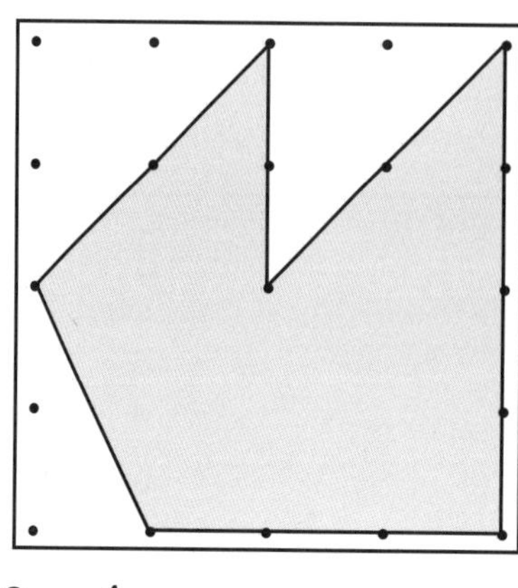

2. $A =$ ____________

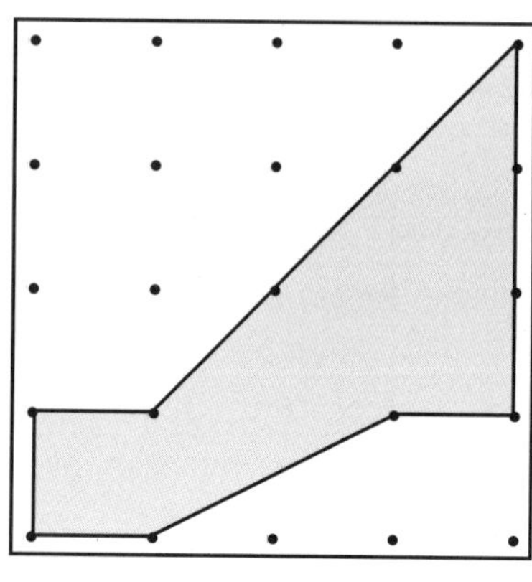

3. $A =$ ____________

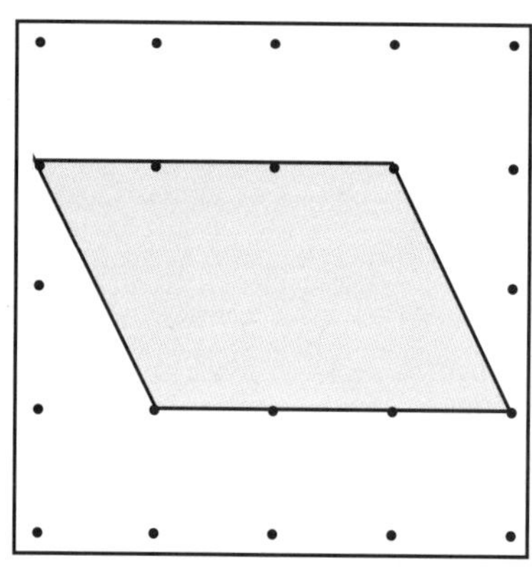

4. $A =$ ____________

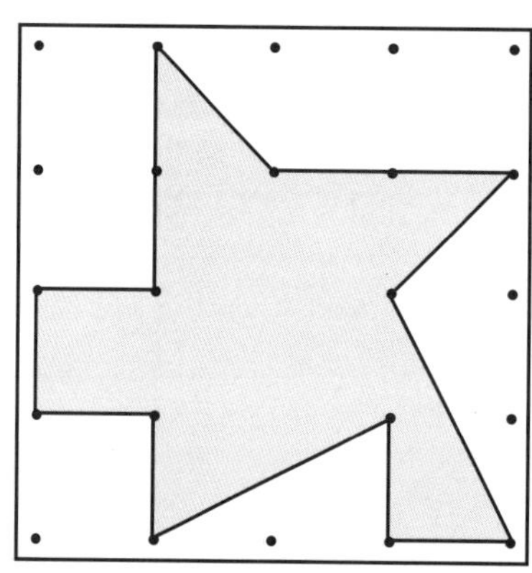

5. $A =$ ____________

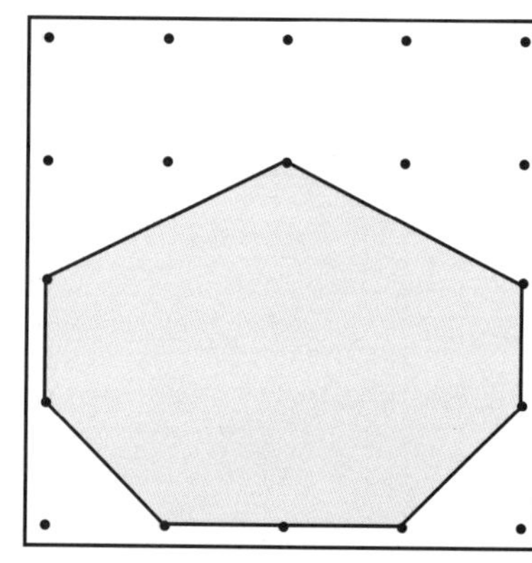

6. $A =$ ____________

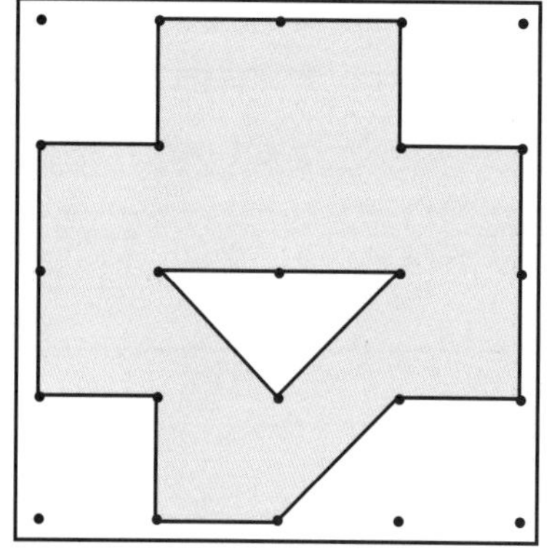

7. $A =$ ____________

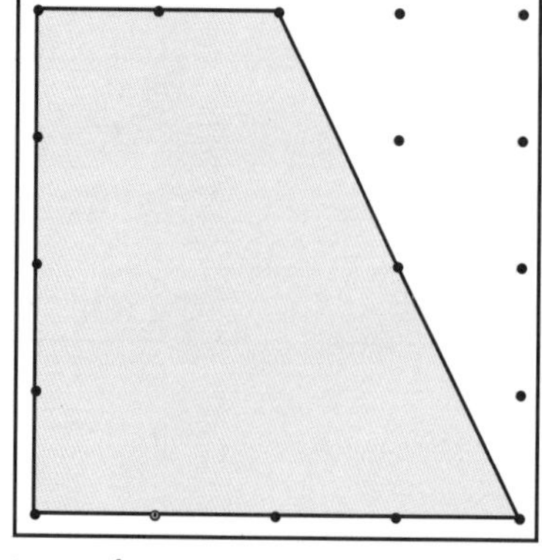

8. $A =$ ____________

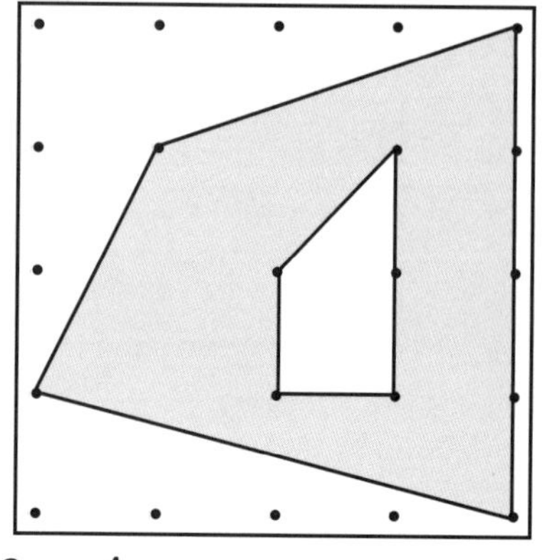

9. $A =$ ____________

# LESSON 18: PUZZLING PARALLELOGRAMS

**Developing a Formula for the Area of a Parallelogram**

## Materials

geoboards and rubber bands—1 per student

geoboard paper (see p. 92)—1 sheet per student

markers in different colors

transparency of Parallelogram for Step 1 (see p. 99)

Puzzling Parallelograms activity sheet—1 per student

*optional:* graph paper; scissors

## Procedure

1. Project the transparency or draw the following parallelogram on the board. Have students construct the parallelogram on their geoboards. Discuss the characteristics of parallelograms: opposite sides equal and parallel. Have students use any techniques they know to find the area. Many will box in the triangles and add the areas of the triangles to that of the rectangle.

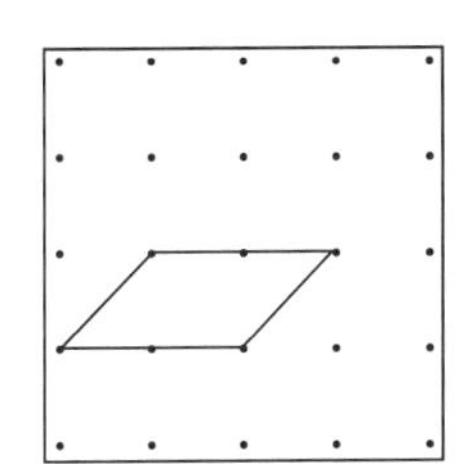

Base = 2

Height = 1

Area = ___

2. Ask: Can you change the parallelogram's shape into a rectangle while keeping the area the same? Discuss strategies and write the base, height, and area of the rectangle on the board. Repeat the procedure with another parallelogram, then ask students to devise a formula for the area of a parallelogram. Most will see that the formula is the same as that of a rectangle of the same base and height. If students have trouble, have them draw parallelograms on graph paper, cut out the triangle at one end of the figure, and slide it over to complete the rectangle.

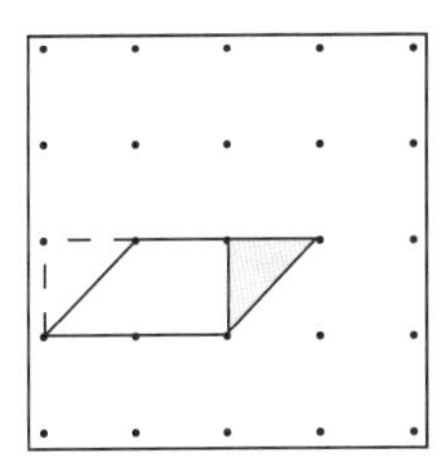

3. Add to the "What We Know About Area" list:
   ***The formula for the area of a parallelogram is $A = b \times h$ or $A = bh$.***

4. Have students use their geoboards to complete the Puzzling Parallelograms activity, drawing their constructions on geoboard paper. Discuss results and strategies.

## Journal Writing

*Finish this sentence:* To do today's geometry activities, I needed the following math skills:

## Extension

- Have students use graph paper to create a 4" by 8" design composed partly of parallelograms. This could be a design unit for a beaded belt. You might want to specify that $\frac{1}{2}$ the area must be parallelograms, or that $\frac{1}{4}$ of the design must be red, and so forth. Share results and strategies.

## Answers

1. Students can start with a rectangle or square with an area of 4 square units and move two corners to make a parallelogram. Bases can be 1 or 2 units; the 5 by 5 geoboard is not big enough to use a base of 4.

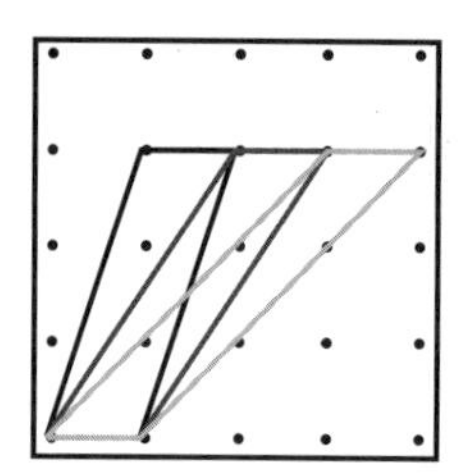
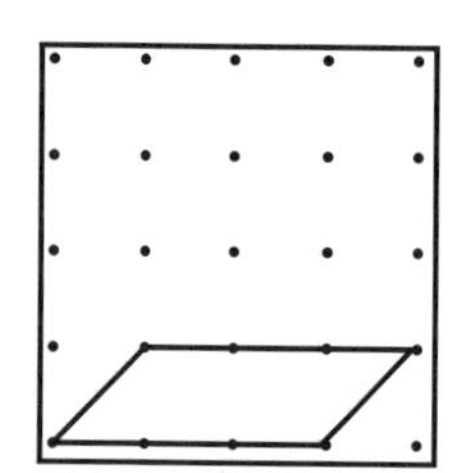

3. $A = 1$ square unit

4. $A = 12$ square units. The largest base and height possible for a rectangle is 4 on this geoboard, but there's not enough room for a parallelogram with a base of 4. A base of 3 is the next smallest.

5. There are four possible parallelograms.

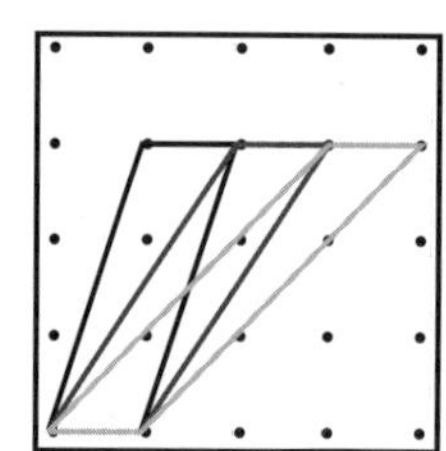
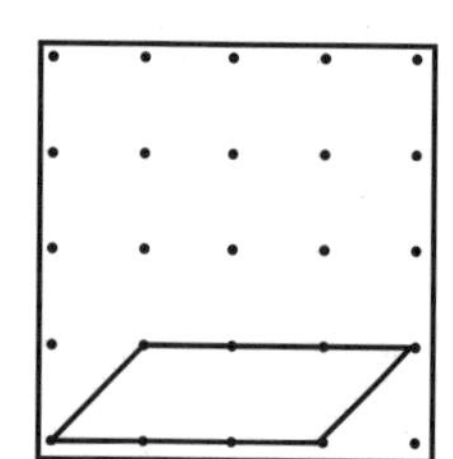

6. There are an infinite number. Students might focus on adding intermediate pins to the geoboard.

Name Date

# PUZZLING PARALLELOGRAMS

1. On your geoboard, construct at least four different nonrectangular parallelograms with an area of 4 square units each. Keep the bases parallel to the edge of the geoboard. Draw your parallelograms on geoboard paper.
2. Explain your thinking as you did this activity.

3. On your geoboard, construct the nonrectangular parallelogram with the smallest area you can.

   $A =$ ______________.

4. Now construct the nonrectangular parallelogram with the largest area you can.

   $A =$ ___________.

   How do you know this parallelogram has the largest area?

5. Construct as many different nonrectangular parallelograms as you can with an area of 3 square units each. Draw these on geoboard paper. How many did you find? ________

6. If you did not have to use a geoboard, how many different noncongruent parallelograms could you draw with an area of 3 square units?

   Why?

# Lesson 19: Birds and Beads

**Finding the Area of All Triangles**

## Materials

geoboards and rubber bands—1 per student

transparency of Isosceles Triangles (see p. 100)

transparency of Scalene Triangles (see p. 101)

Birds and Beads activity sheet—1 per student

*optional:* graph paper, straightedges, and scissors may be used instead of geoboards

## Procedure

*Note:* Students need to know how to "box in" a triangle on the geoboard as shown in Lesson 17. They should also know that the area of a parallelogram is $A = bh$.

1. Discuss what students already know about finding the area of right triangles, including cutting rectangles apart on the diagonal and boxing in triangles on the geoboard to construct a rectangle. The formula $A = \frac{1}{2}bh$ works for right triangles. Ask: Does it work for all triangles? What other kinds of triangles need to be investigated?

2. Project Figure A from the Isosceles Triangles transparency. Have students make Figure A (an isosceles triangle, with two equal angles) on their geoboards. Have them use what they already know to find its area. You may have to hint about separate areas for separate parts of the triangle (see Figure B).

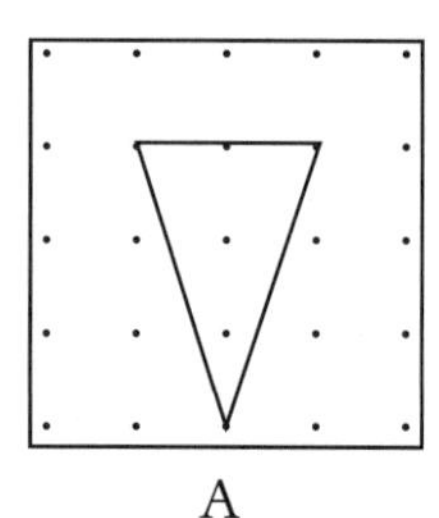
A

B

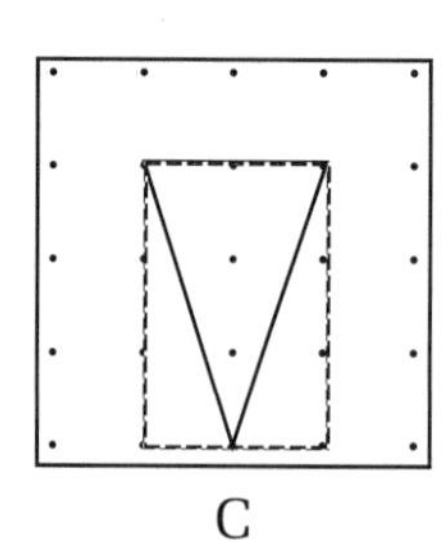
C

Discuss students' strategies and results.

3. Project Figure C. Have students make Figure C on their geoboards, an isosceles triangle enclosed in a rectangle. Can they find another way to show that the formula $A = \frac{1}{2}bh$ works? If they don't think of subtracting the extra areas, you may have to hint about finding the areas around the main triangle.

Add to the "What We Know About Area" list:

***The formula for the area of an isosceles triangle is $A = \frac{1}{2}bh$.***

4. Project the Scalene Triangles transparency, revealing the figures one at a time, and have students investigate the area of scalene triangles (those with no equal angles) by making them on their geoboards.

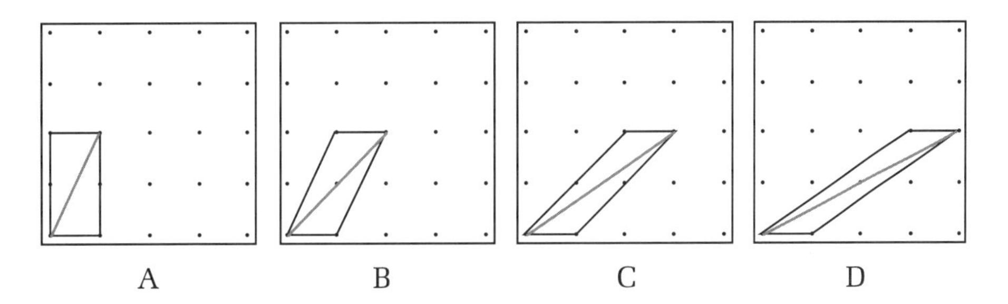


For each figure, ask:

- What is the base? What is the height?
- Are the triangles congruent? How do you know?
- What is the area of the rectangle (or parallelogram)?
- What is the area of one of the triangles?
- Does this fit the formula?

Add to the "What We Know About Area" list:

***The formula for the area of a scalene triangle is $A = \frac{1}{2}bh$. This formula works for all triangles.***

5. Have students complete the Birds and Beads activity sheet. Discuss results and strategies.

## Journal Writing

In what ways are triangles, rectangles, and parallelograms related?

## Extensions

- Have students use graph paper and colored markers to design very small wall hangings or belts with birds and triangles of specific areas in the design. Research simple weaving techniques, and have students weave their wall hangings.
- Have students use their knowledge of area to investigate probability. Each student will need a sheet of graph paper to use as a game board. The outline boundary of the grid is the perimeter of the game board. Have students use straightedges to draw triangles: the base is the same as one side of the grid, and the apex is on the side of the grid parallel to the base.

There will be a variety of triangles. Ask: What is the area of each triangle? What is the area of each game board? If no skill were involved and a counter landed on the game board, what is the chance it would land inside the triangle? (*50%, or 1 out of 2*)

- Have students devise a method for testing this theory. Will some students' results be different from others? Why? Ask students to predict the effect of practicing on the results. Have students practice hitting the triangle by tossing 20 counters. Have them test the probability of landing a counter in the triangle after practicing. Did their scores increase?

## Answers

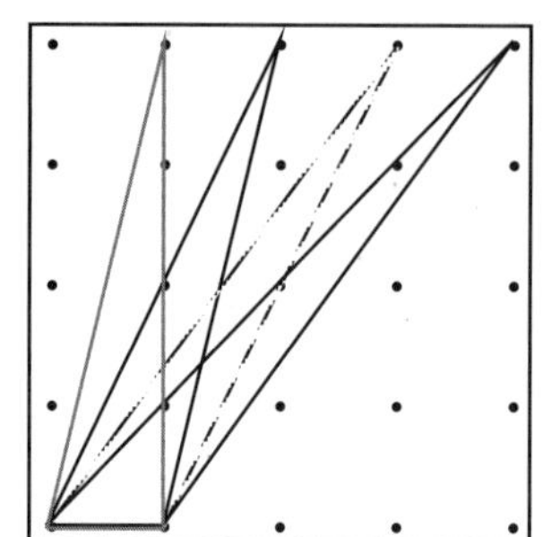
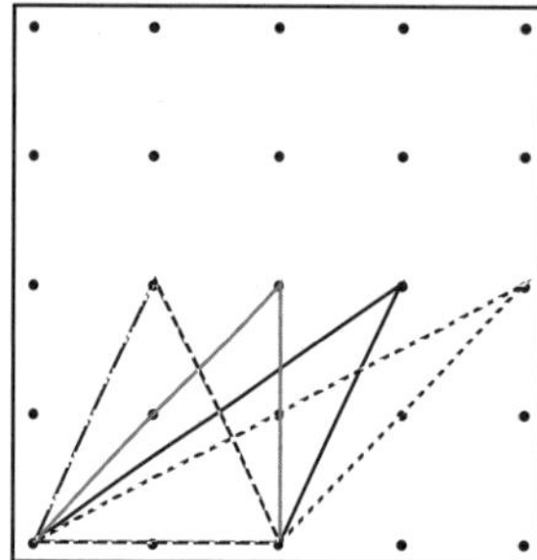
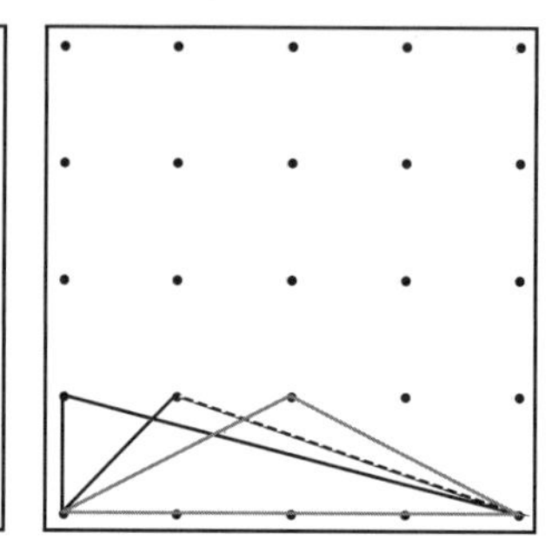

Name Date

# BIRDS AND BEADS

Hans is weaving beads to make a belt. He plans to make a center triangle design with a bird on either side of the triangle. The center triangle will be made of red beads. Hans has enough red beads to cover 2 square inches. What kinds of triangles can he make?

Use your geoboard to find as many different triangles as you can with an area of 2 square units. Keep the bases of your triangles parallel to an edge of the geoboard. Draw your triangles on the geoboards below. You will have to draw more than one triangle on each geoboard.

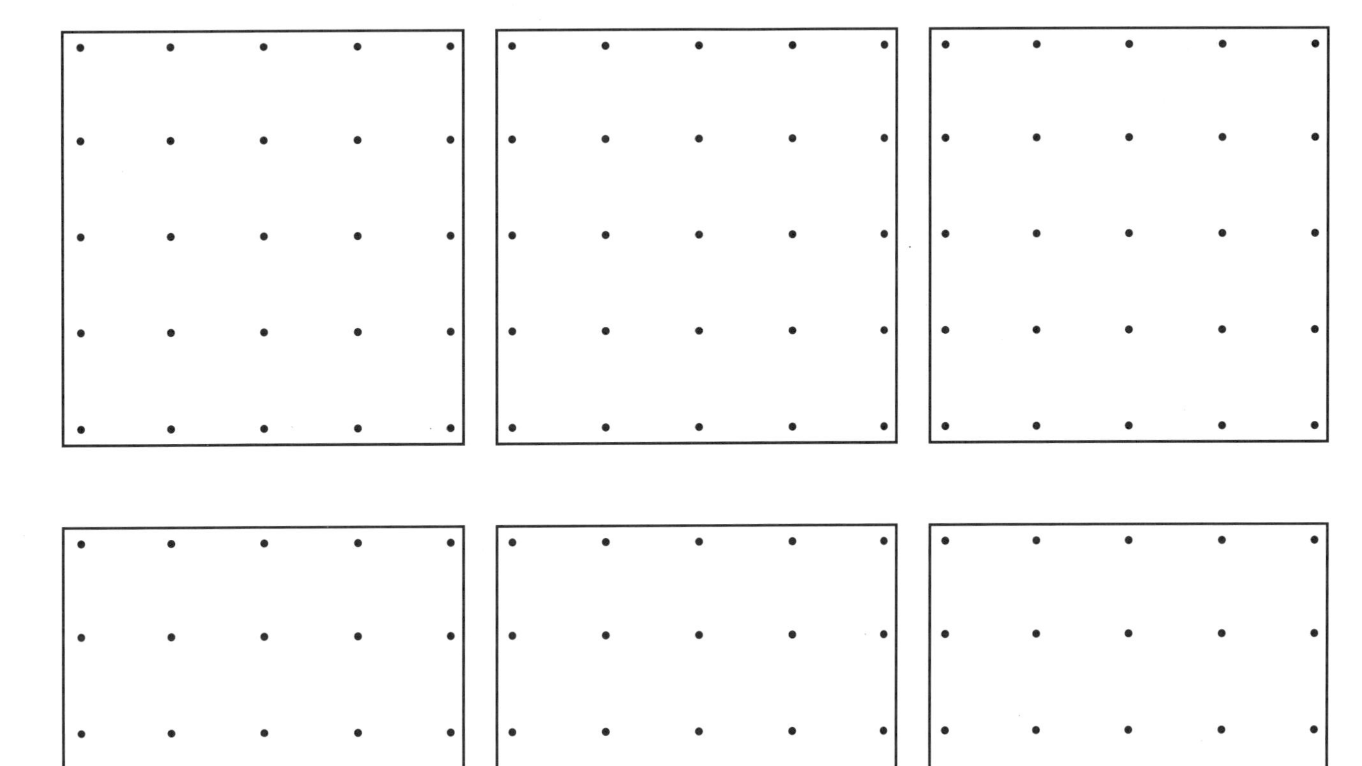

How many *different* triangles did you find? ________. What was your strategy?

On the back of this sheet, tell why you think you have found them all.

# LESSON 20: TASTY TRAPEZOIDS

**Developing a Formula for the Area of a Trapezoid**

## Materials

1-cm graph paper (see p. 96)—1 sheet per student

scissors

Tasty Trapezoids activity sheet, pages 1 to 2—1 set per student

## Procedure

1. On their graph paper, have students draw a trapezoid with a bottom base 8 units long, top base 3 units long, and height of 2 units. There are several possible shapes; here is one:

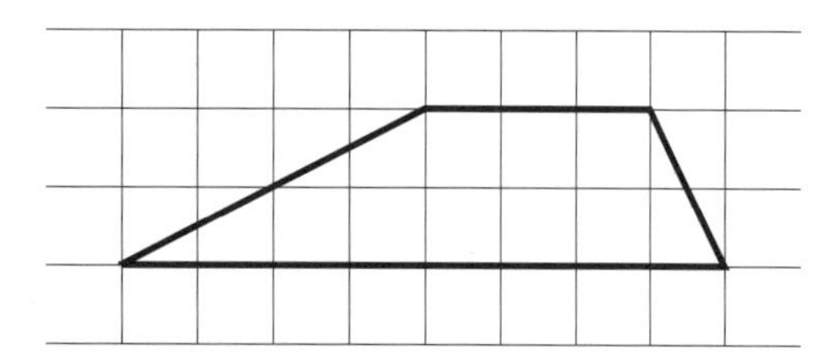

Have students find the area of their trapezoids by any method they choose. Discuss strategies and results:

- Were there differences in area among the different shapes?
- You already know quite a bit about finding area. How did you use your knowledge to solve this problem?
- Did anyone devise a way that will work for all trapezoids?

2. Remind students that to find the formula for the area of a triangle, they boxed it in to create a rectangle. In effect, they constructed a new triangle, identical to the first, and created a figure from the two whose area they knew how to find. Ask: Can a similar approach be used to find the formula for the area of a trapezoid?

Suggest that students create a second trapezoid identical to the first, and label the top bases $b_1$, the bottom bases $b_2$, and the heights $h$. Labels should be inside the figures:

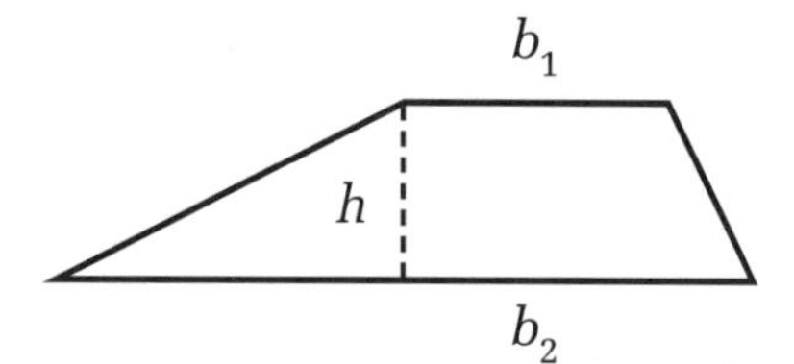

Students should cut out both trapezoids and try to construct a familiar figure for which they know the area formula. They should be able to see that they can make a parallelogram whose area is $h \times (b_1 + b_2)$ or $2 \times (3 + 8) = 22$ square units:

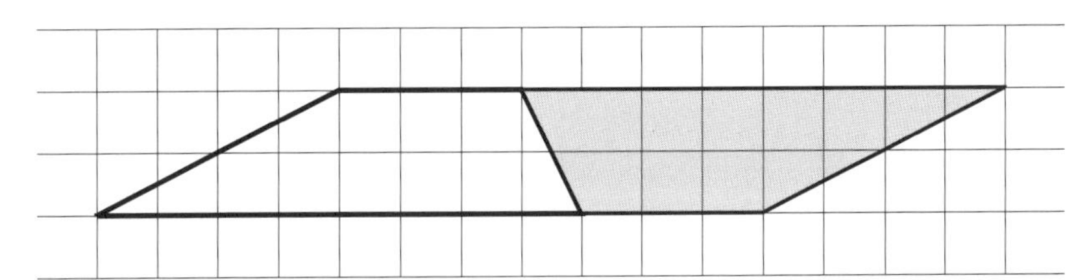

The trapezoids are congruent, and each is a half of the parallelogram. They can use half the formula for a parallelogram's area. The formula can be written $\frac{1}{2}$ × height × (base 1 × base 2) or $\frac{1}{2}h(b_1 + b_2)$. Ask: Does this method work for all the trapezoids?

3. Have students try this method with trapezoids of their choosing, making sure to include some trapezoids that have an oblique angle between the bottom base and a side. Students should chart their results for $h$, $b_1$, $b_2$, and area:

| $h$ | $b_1$ | $b_2$ | Area = $\frac{1}{2}h(b_1 + b_2)$ |
|---|---|---|---|
| | | | |
| | | | |
| | | | |
| | | | |

4. Compare results. Add to the "What We Know About Area" list:
***The formula for area of a trapezoid is*** $A = \frac{1}{2}h(b_1 + b_2)$.

5. Work through the directions for the Tasty Trapezoids activity with students. Many will not realize that they can find half the height or half the sum of the bases to figure the area of a trapezoid.

6. Have students compete the Tasty Trapezoids activity. Discuss strategies, results, and any patterns they noticed.

## Journal Writing

Imagine you're a trapezoid. Write a letter to a rectangle.

*Answer this question:* Where do you find trapezoids in the world?

## Extensions

- Plan a geometry party with your students. Discuss what kinds of refreshments come in geometric shapes.
- Make trapezoid cookies, or construct them from clay or construction paper. Make a bulletin board display from the paper ones.
- Do trapezoids tessellate? Use pattern blocks (see pp. 84–87) or isometric graph paper (see p. 88) to investigate possible tessellations of trapezoids and combinations of other shapes. See pp. 89–91 for tessellation masters.

## Answers

4. There are 11 trapezoids of area 4 that satisfy the given conditions.

5. Students may see that they can find all the trapezoids by moving the top base over one square each time. The longer the bottom base, the more positions possible for the top base (going left to right) before it reaches the center position. As it moves to the right of the center position, the trapezoids formed are congruent to ones already made.

   Students should notice that as the height increases, the total for the combination of bases decreases. Larger numbers have a greater number of combinations possible for $b_1 + b_2$.

| $h$ | $b_1 + b_2$ | $b_1 + b_2$ | Different Trapezoids Possible |
|---|---|---|---|
| 1 | 8 | 1 + 7 | 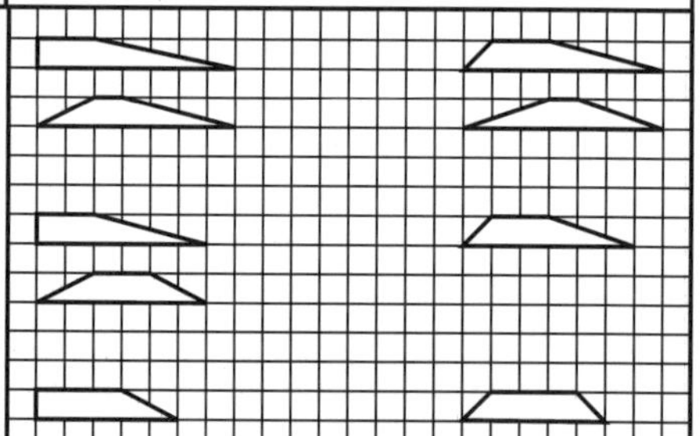 |
| | | 2 + 6 | |
| | | 3 + 5 | |
| | | 4 + 4 | makes a rectangle |
| 2 | 4 | 1 + 3 |  |
| | | 2 + 2 | makes a square |
| 4 | 2 | 1 + 1 | makes a rectangle |
| 8 | 1 | 1 + 0 | makes a triangle |

Name Date

# TASTY TRAPEZOIDS, PAGE 1

Mr. Jordan's class is having a geometry party. Tran and Jamie have volunteered to make trapezoid-shaped sugar cookies. They will roll the dough until it is thin, place a grid over it, and cut out the cookies. Each spoonful of dough makes 4 square inches of cookie.

They decide that each cookie should have an area of 4 square inches, and they'll try to have as many trapezoid shapes as possible with this area. They also decide that the outside angle between the bottom base and the sides should not be less than 90°:

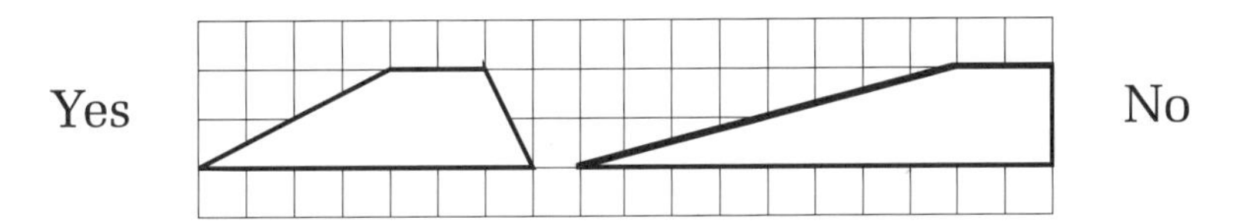


How many ways can Tran and Jamie cut the cookies—how many trapezoid shapes do you think are possible with an area of 4 square inches? Remember that a cookie is the same shape even if it is turned over.

Remember that the formula for the area of a trapezoid is $A = \frac{1}{2}h(b_1 + b_2)$. A must equal 4 for Tran and Jamie, so $h(b_1 + b_2)$ must equal 8. Also:

- It helps to be organized. Start with $h = 1$ inch.
- How many whole-number ways can $(b_1 + b_2)$ equal 8? Here's a start:

| *Height* | $b_1 + b_2$ | $b_1 + b_2$ *Combinations* | **Different Trapezoids Possible** |
|---|---|---|---|
| 1 | 8 | 1 + 7 | 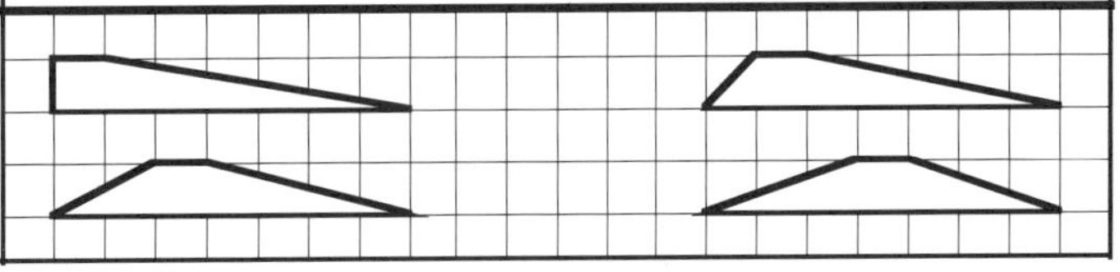 |

Use the table on page 2. Fill it in, and draw each shape.

- Watch out for shapes that are the same except for being flipped.
- Be careful—*some combinations of heights and bases don't make trapezoids!* For those, write what shape they do make. Example: If $h = 1$, and the $b_1$, $b_2$ combination is 4 + 4, a rectangle is formed.

Name Date

# TASTY TRAPEZOIDS, PAGE 2

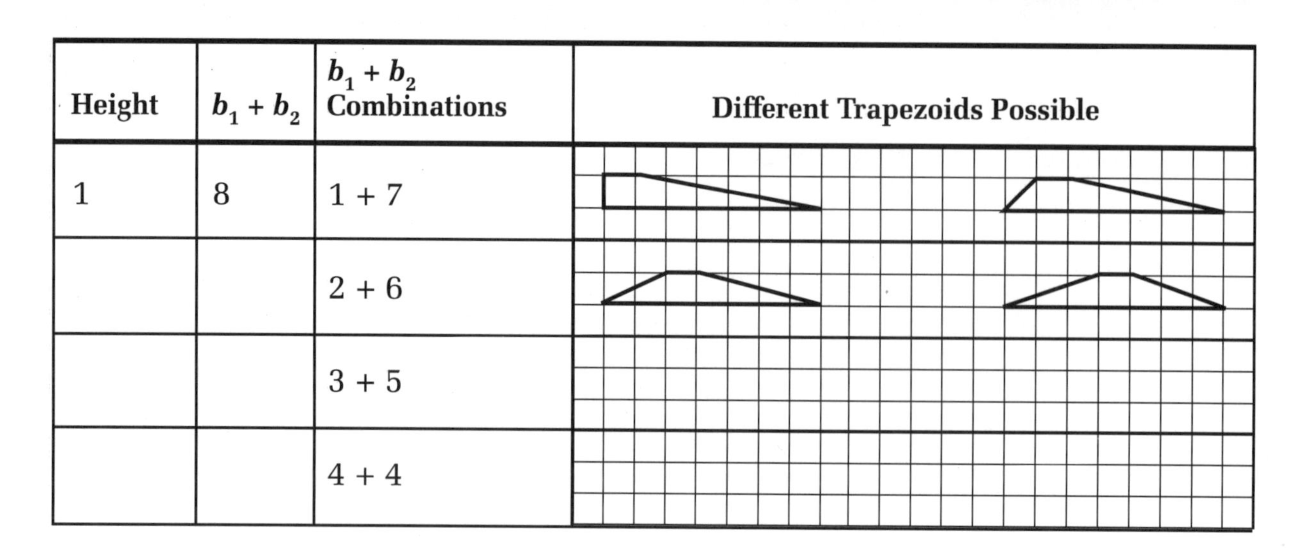

| Height | $b_1 + b_2$ | $b_1 + b_2$ Combinations | Different Trapezoids Possible |
|---|---|---|---|
| 1 | 8 | 1 + 7 | |
| | | 2 + 6 | |
| | | 3 + 5 | |
| | | 4 + 4 | |

Now try other heights. Remember, $h(b_1 + b_2)$ must equal 8.

| Height | $b_1 + b_2$ | $b_1 + b_2$ Combinations | Different Trapezoids Possible |
|---|---|---|---|
| 2 | 4 | | |
| | | | |
| 4 | | | |
| 8 | | | |

4. The total number of different trapezoids with an area of 4 square inches Tran and Jamie can cut out is ______________.

5. How did you find all the trapezoids?

# Lesson 21: Pizza Time

### Developing a Formula for the Area of a Circle

## Materials

calculators

transparent tape or glue stick

$\frac{1}{4}$-in. or 1-cm graph paper (see pp. 93 and 96)—1 sheet per student

rulers

scissors

writing paper—1 sheet per student

transparency of Kepler's Method for Finding the Area of a Circle (see p. 102)

Pizza Time activity sheet, pages 1 to 2—1 set per student

## Procedure

1. Explain the following scenario: Everyone likes the pizzas José and Anya make. José makes wonderful sauce, and Anya's dough is just right. José and Anya know their pizza is popular, and there's no pizzeria in their neighborhood. They've decided to go into business as Anjo's Pizzas. Now they need to draw up a business plan. Discuss with students:
   - What expenses will José and Anya need to consider? (*ingredients for the dough and sauce; other ingredients such as cheese and mushrooms; advertising; utilities such as the phone, and heat for the oven; labor if they pay someone to do deliveries*)
   - How should José and Anya decide the price for their pizzas? (*They will have to estimate sales, figure the cost to them for each pizza, consider what other companies charge for their pizza, and make sure they make a profit.*)

2. Each ounce of José's sauce will cover 10 square inches. José and Anya start by finding the area of their individual-size pizza, which is 7" in diameter. Discuss what facts students already know about circles. List these facts on the board, including $d = 2r$ and $C = 2pr$.

3. Have students complete page 1 of the Pizza Time activity, using any materials they wish.

4. Discuss students' results and methods:
   - Were some more accurate than others?
   - Will anyone's method work for any circle?

5. Have students complete page 2 of the Pizza Time activity.

6. Discuss results and methods. If students have not devised a method similar to the following, devised by astronomer Johannes Kepler (1571–1630) in 1609, project the transparency of Kepler's Method for Finding the Area of a Circle (see page 102). Suggest that students try Kepler's method. Students cut apart their circle on the lines, reassemble the parts into a more familiar figure, and find the area of that figure. Ask:
   - How accurate do you think your estimates are?
   - Did anyone use a figure other than a parallelogram?

   Kepler's method says the area of the parallelogram approximates the area of the circle:

   $A = \pi r \times r$

   $A = \pi r^2$

   So, the area of a 7" pizza is:

   $A = \pi \times 3.52$

   $A = 3.14 \times 12.25$

   $A = 38.46$ sq. in.

   Since each ounce of José's sauce will cover 10 square inches, they will need 3.8 ounces.

7. Add to the "What We Know About Area" list:
   ***The area of a circle can be found using the formula $A = \pi r^2$.***

8. Have students figure out the amount of sauce needed for 9", 14", and 18" pizzas (*25.4 ounces, 61.6 ounces, 101.8 ounces*).

## Journal Writing

Write a problem about pizzas or pizza ingredients that requires finding the area of a circle.

## Extensions

- Find the area of a square of very heavy paper or cardboard, then weigh the square. Construct the largest circle possible on it, cut out the circle, and weigh the circle. What part of the original weight is now contained in the circle? What part of the original area is now contained in the circle? What is the area of the circle? Point out that students can use their knowledge of ratios to find the area this way. How do the results compare with those obtained by using a formula?
- Give students a 12" piece of twine to tie into a loop. Have them enclose different figures with it and find the areas. What figures give the largest area for the perimeter? (*circles*)

Name Date

# PIZZA TIME, PAGE 1

Can you find the area of this pizza? Its diameter is 7 inches.

On a separate sheet of paper, write your answer and what you did to find it.

Name Date

## PIZZA TIME, PAGE 2

Can you find a way to use scissors and what you already know about circles to find the area of this pizza? Its diameter is 7 inches.

On a separate sheet of paper, write your answer and what you did to find it. Also write a formula for your method that would work for all circles.

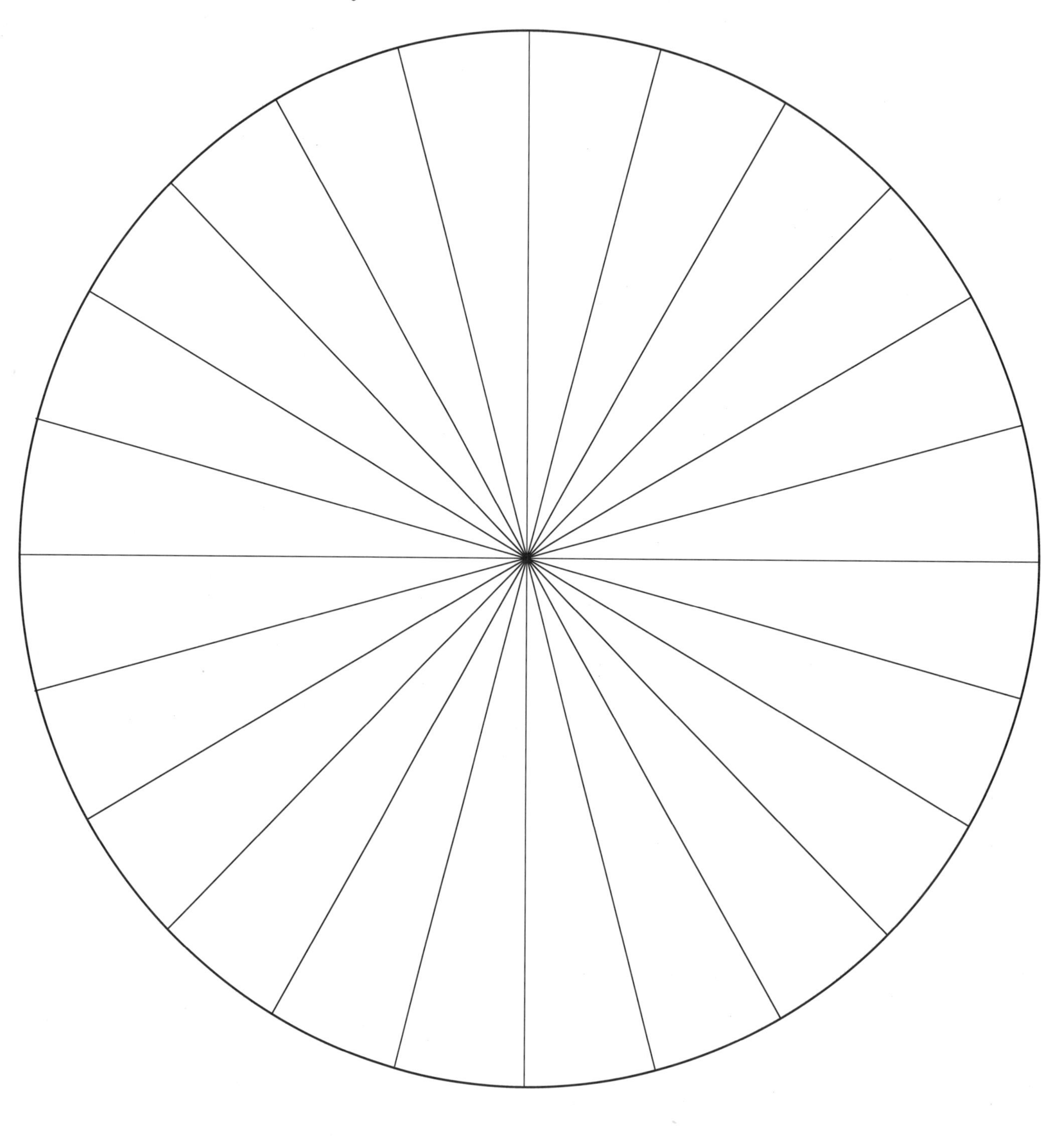

# BLACKLINE MASTERS

## Understanding the Problem

**0** Completely misinterprets the problem.

**1** Misinterprets part of the problem.

**2** Completely understands the problem.

## Choosing and Implementing a Solution Strategy

**0** Makes no attempt or uses a totally inappropriate strategy.

**1** Chooses a partly correct strategy based on interpreting part of the problem correctly.

**2** Chooses a strategy that could lead to a correct solution if used without error.

## Getting the Answer

**0** Gets no answer or a wrong answer based on an inappropriate solution strategy.

**1** Makes copying error or computational error, gets partial answer for a problem with multiple answers, or labels answer incorrectly.

**2** Gets correct solution.

# CLASS PROBLEM-SOLVING CHECKLIST

Dates ______ ______ ______ ______ ______

| Names | Understand Problem | Choose/Use Strategy | Get Answer | Understand Problem | Choose/Use Strategy | Get Answer | Understand Problem | Choose/Use Strategy | Get Answer | Understand Problem | Choose/Use Strategy | Get Answer | Understand Problem | Choose/Use Strategy | Get Answer |
|---|---|---|---|---|---|---|---|---|---|---|---|---|---|---|---|
| | | | | | | | | | | | | | | | |
| | | | | | | | | | | | | | | | |
| | | | | | | | | | | | | | | | |
| | | | | | | | | | | | | | | | |
| | | | | | | | | | | | | | | | |
| | | | | | | | | | | | | | | | |
| | | | | | | | | | | | | | | | |
| | | | | | | | | | | | | | | | |
| | | | | | | | | | | | | | | | |
| | | | | | | | | | | | | | | | |
| | | | | | | | | | | | | | | | |
| | | | | | | | | | | | | | | | |
| | | | | | | | | | | | | | | | |
| | | | | | | | | | | | | | | | |
| | | | | | | | | | | | | | | | |
| | | | | | | | | | | | | | | | |
| | | | | | | | | | | | | | | | |
| | | | | | | | | | | | | | | | |
| | | | | | | | | | | | | | | | |
| | | | | | | | | | | | | | | | |
| | | | | | | | | | | | | | | | |
| | | | | | | | | | | | | | | | |
| | | | | | | | | | | | | | | | |
| | | | | | | | | | | | | | | | |
| | | | | | | | | | | | | | | | |

# PATTERN BLOCK HEXAGONS

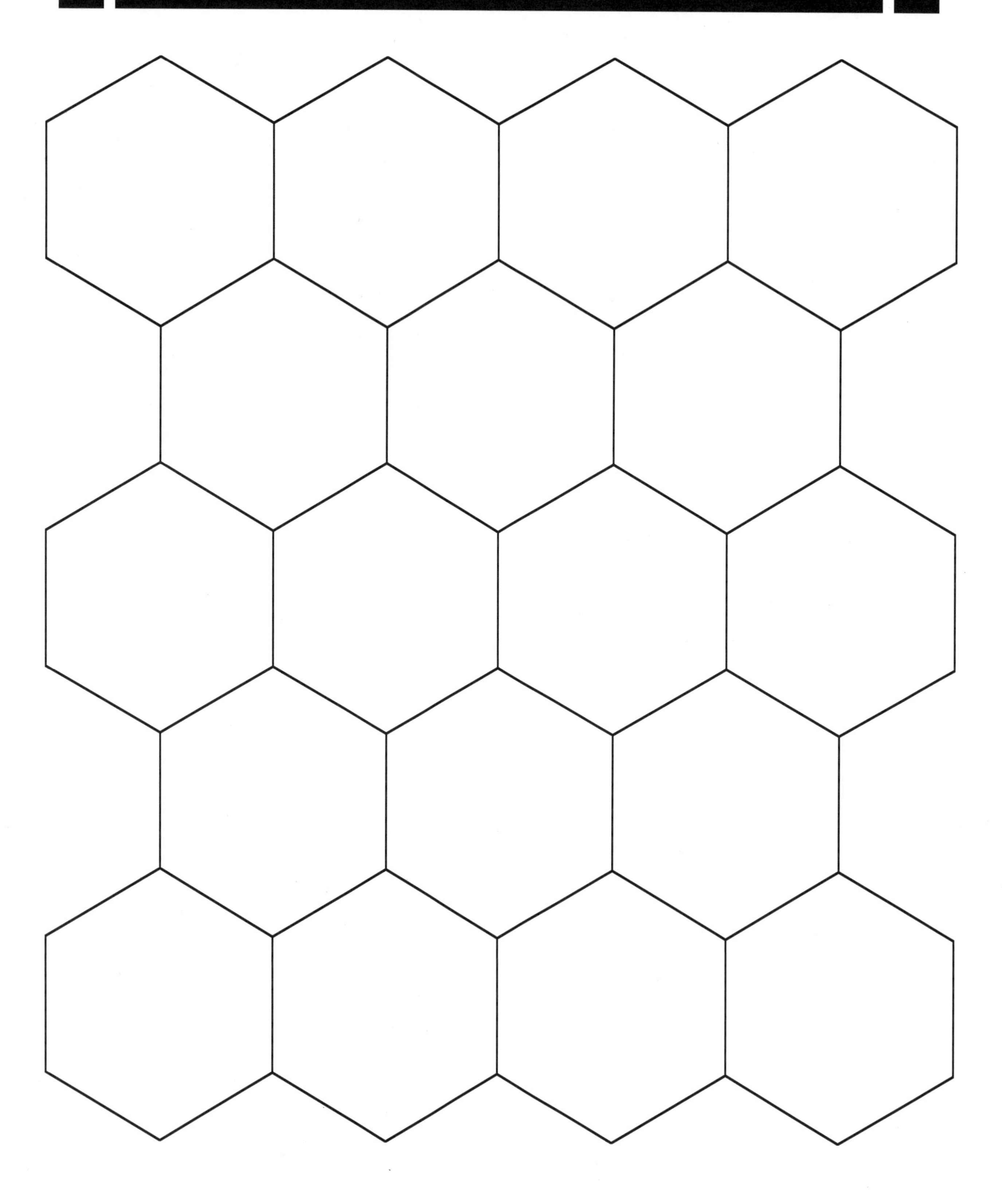

# PATTERN BLOCK RHOMBUSES

# PATTERN BLOCK TRAPEZOIDS

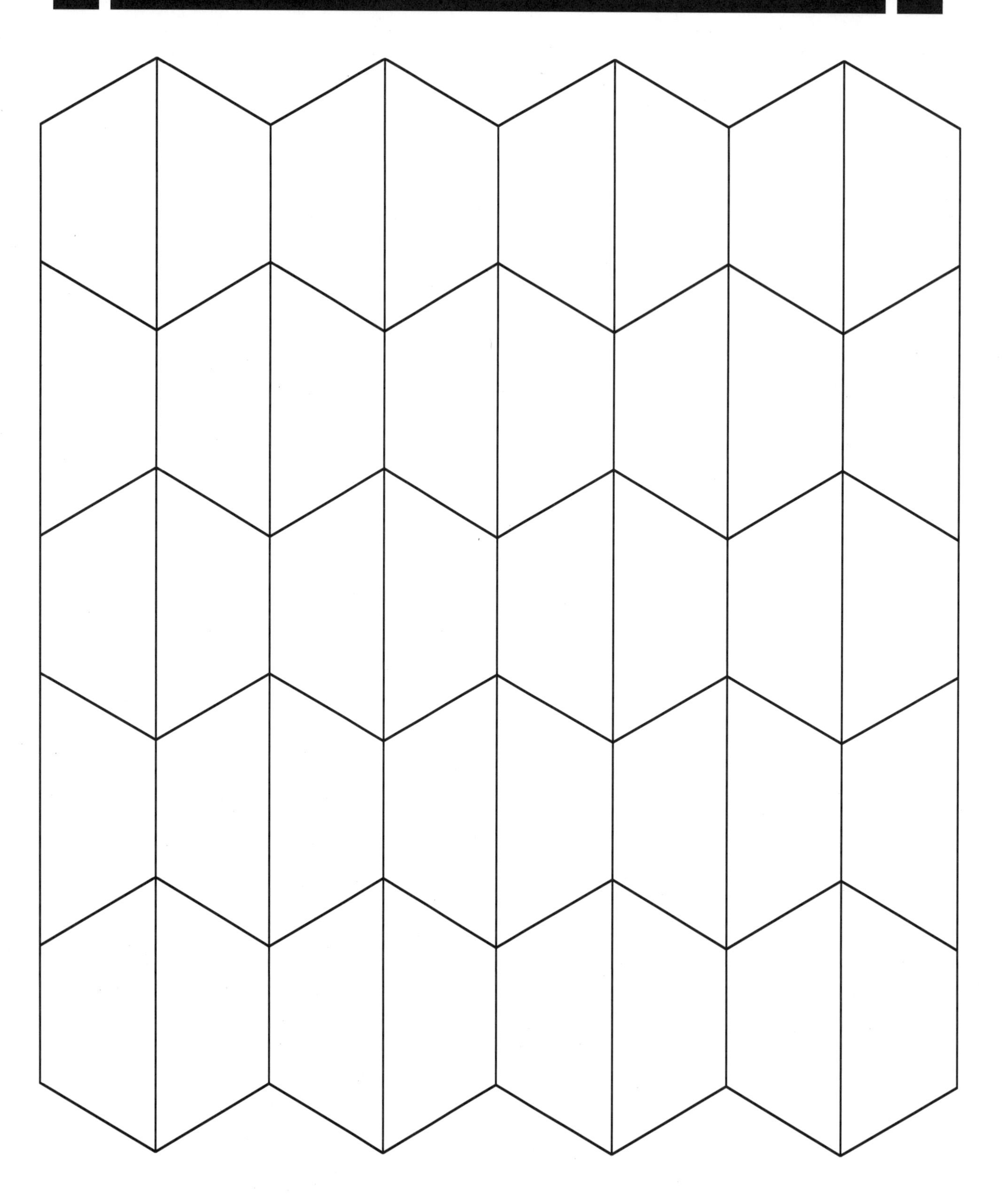

# PATTERN BLOCK TRIANGLES

# ISOMETRIC GRAPH PAPER

# TESSELLATION MASTER 1

# TESSELLATION MASTER 2

# TESSELLATION MASTER 3

# GEOBOARD PAPER

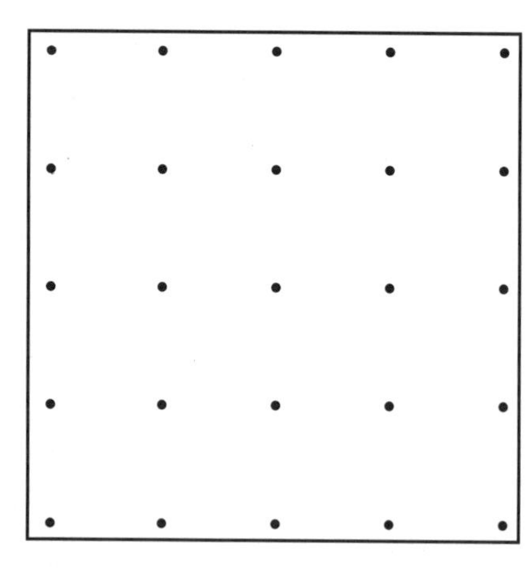

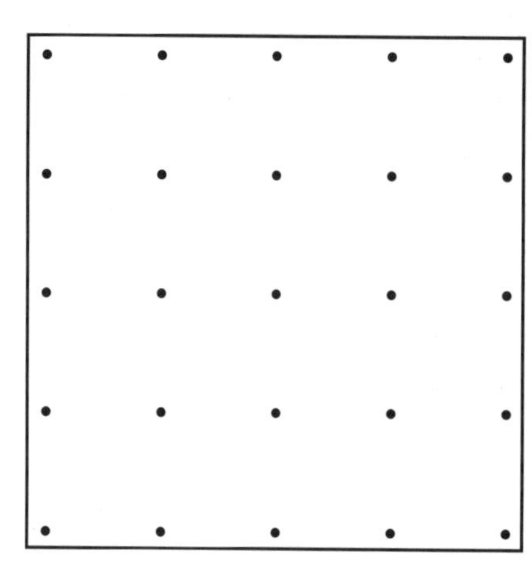

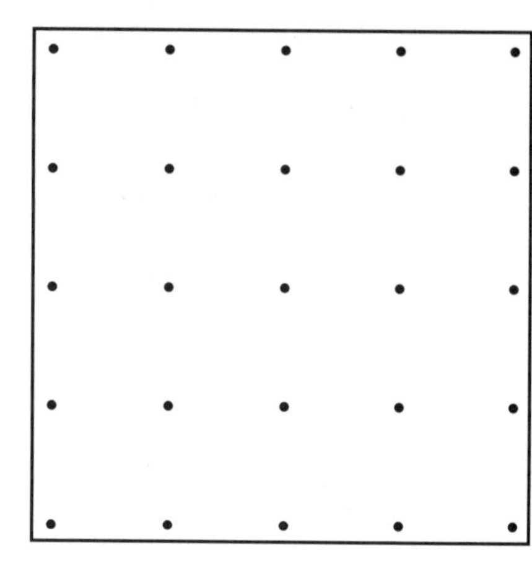

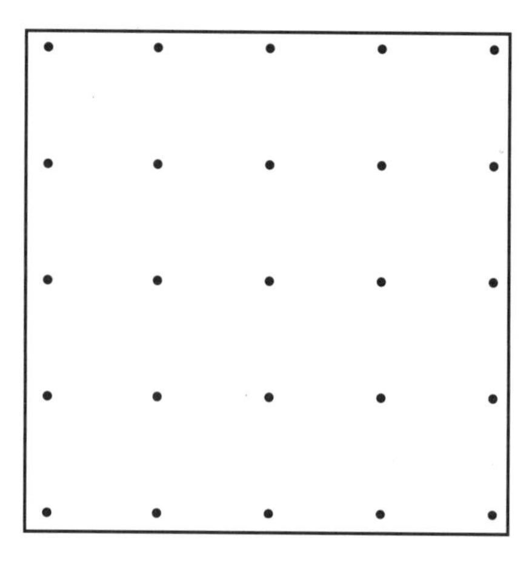

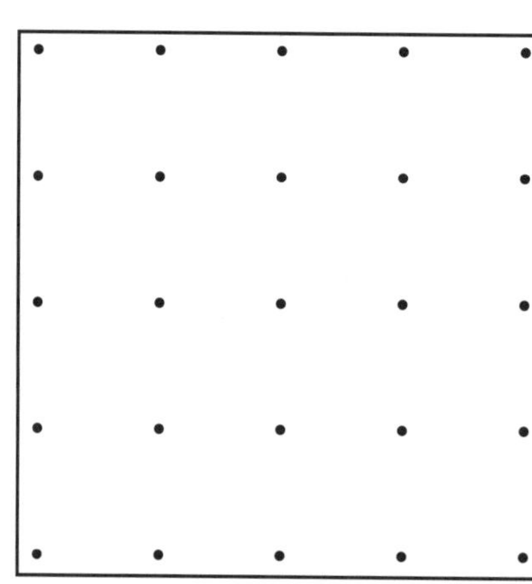

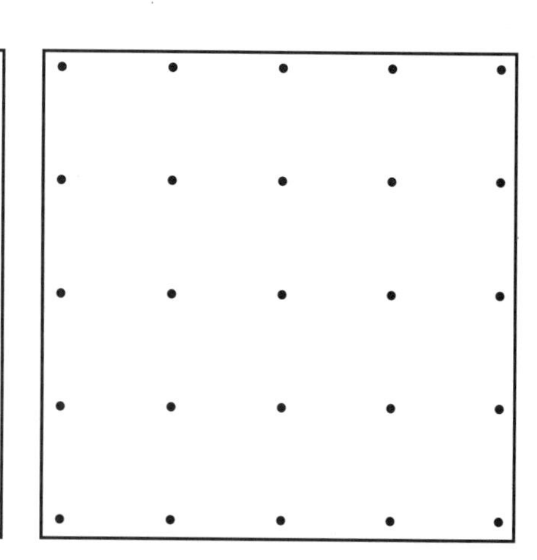

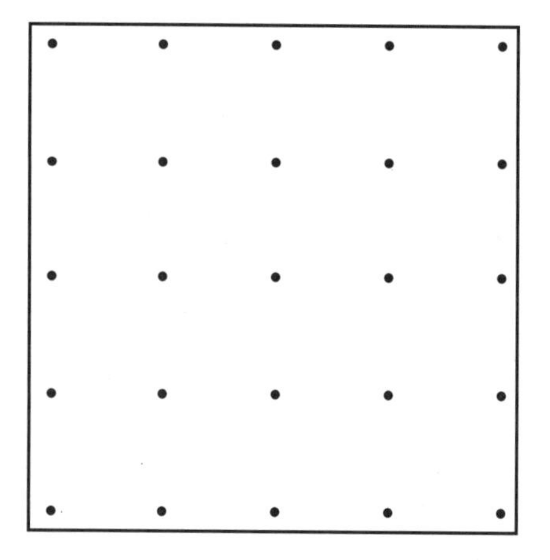

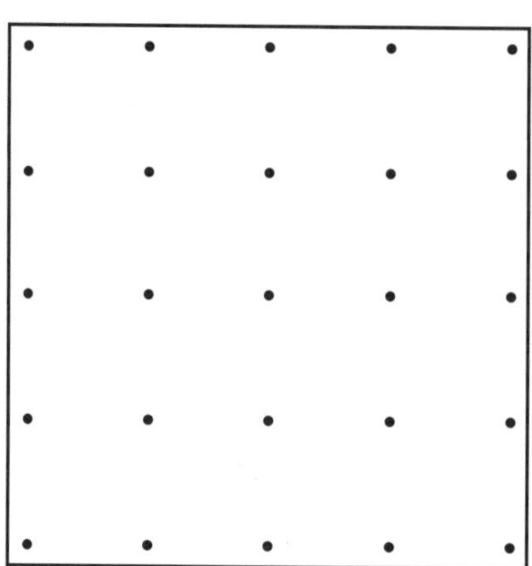

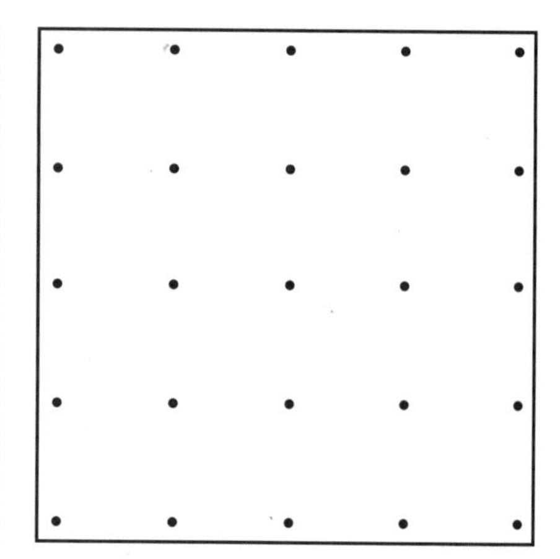

# $\frac{1}{4}$-INCH GRAPH PAPER

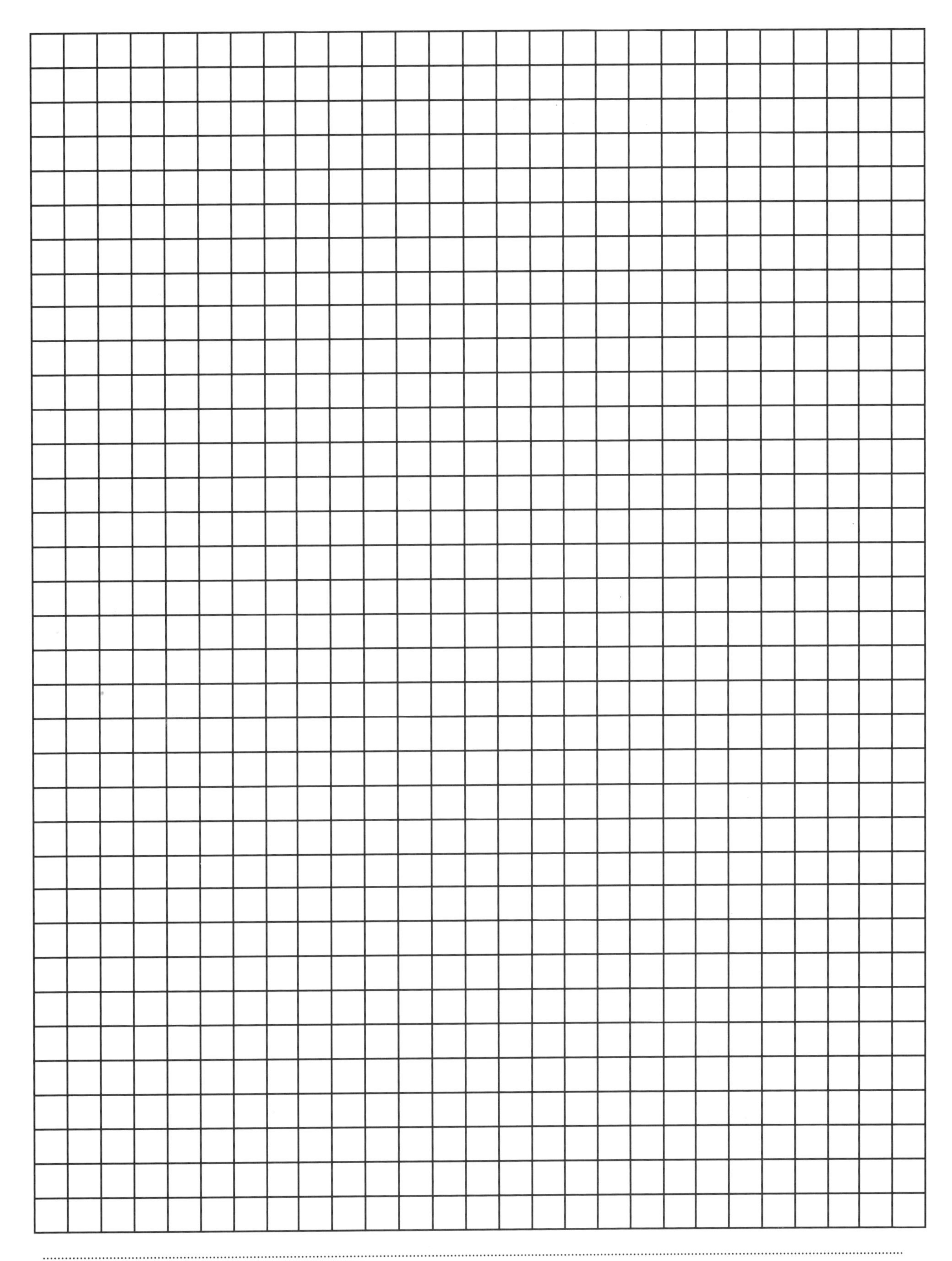

# $\frac{1}{2}$-INCH GRAPH PAPER

# 1-INCH GRAPH PAPER

# 1-CM GRAPH PAPER

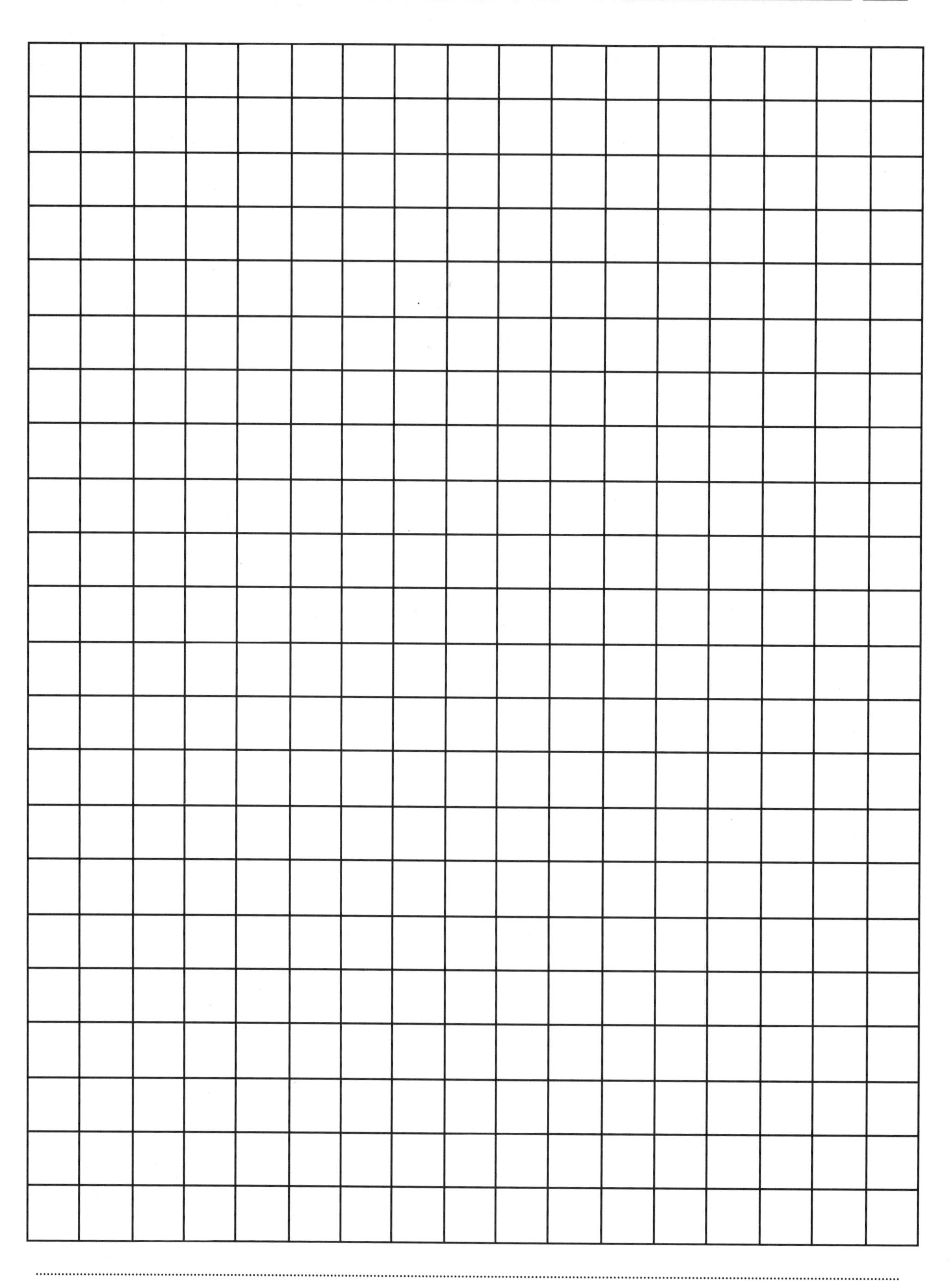

9'

3'

4'

10'

3'

6'

12'

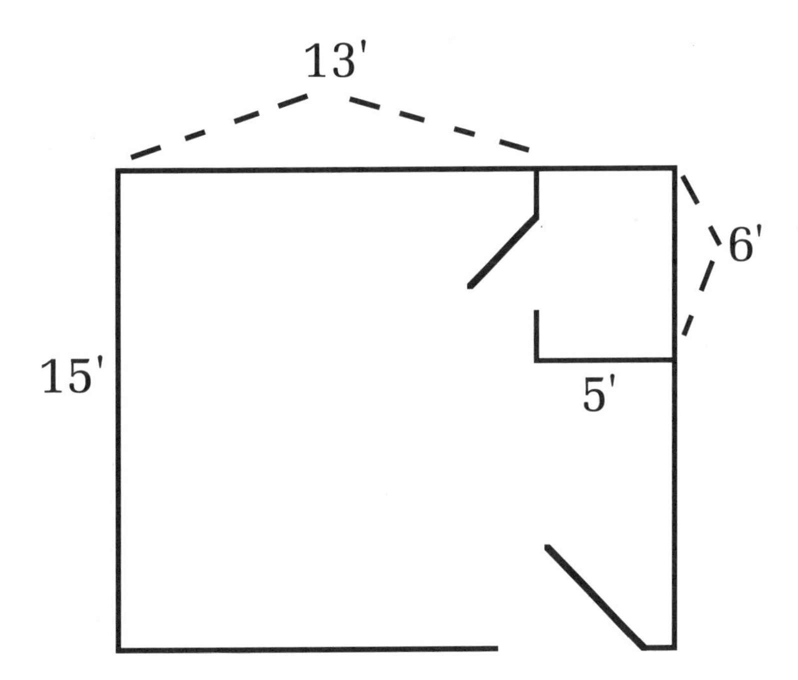

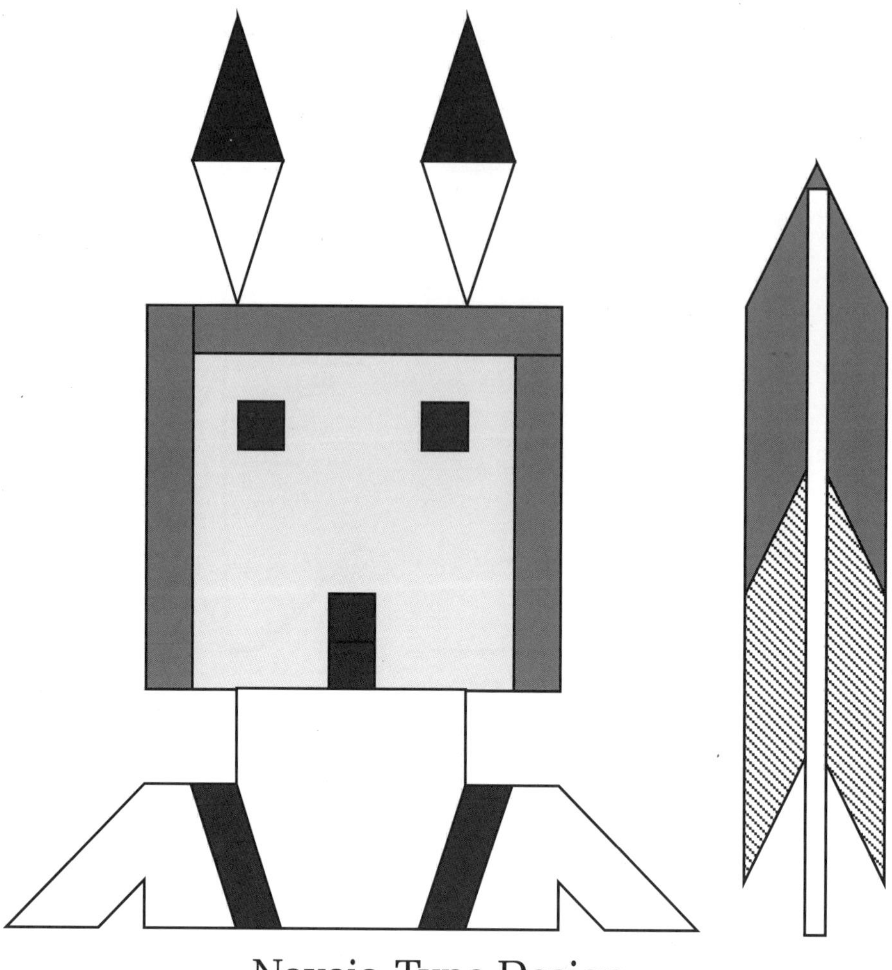

Navajo-Type Design

Seminole-Type Design

# PARALLELOGRAM FOR STEP 1

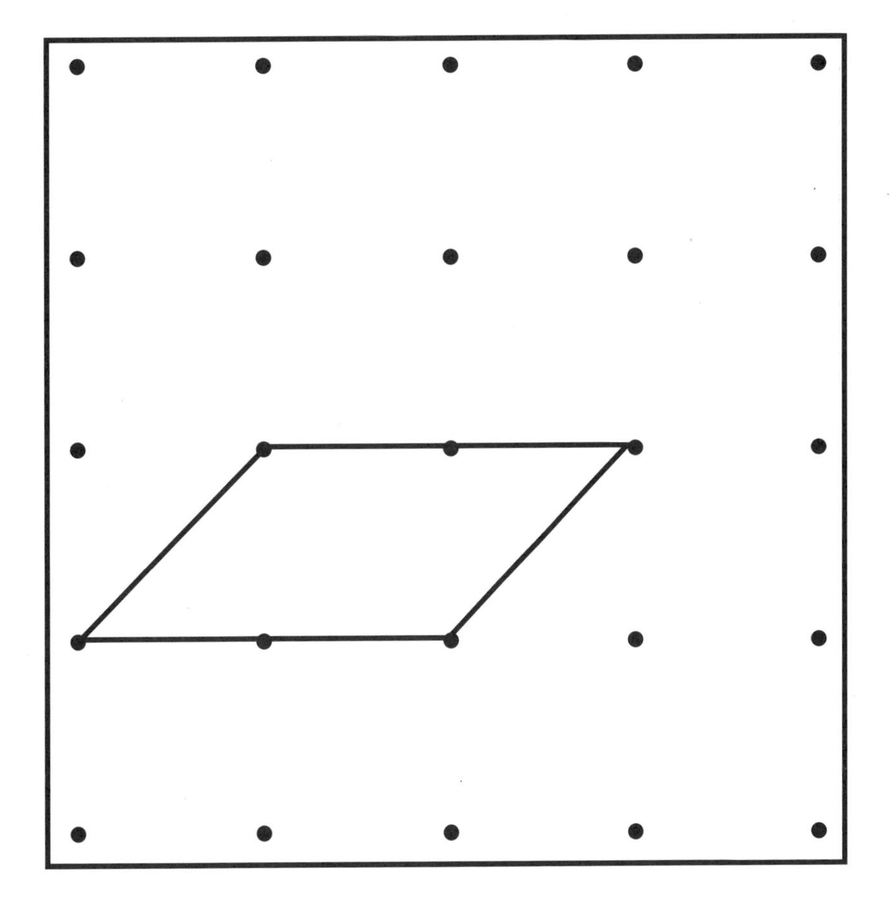

Base = 2
Height = 1
Area =

# ISOCELES TRIANGLES

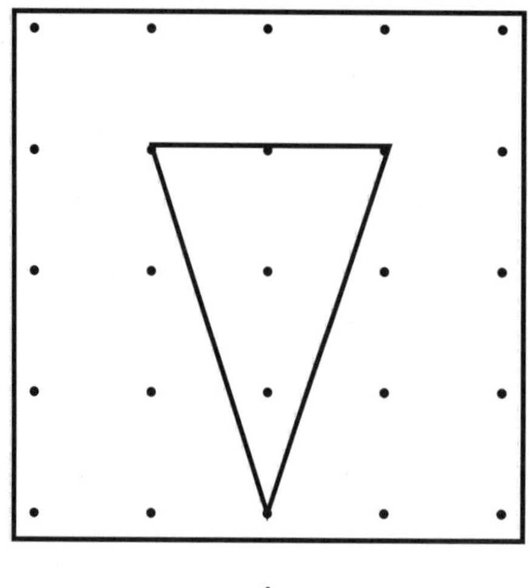

A

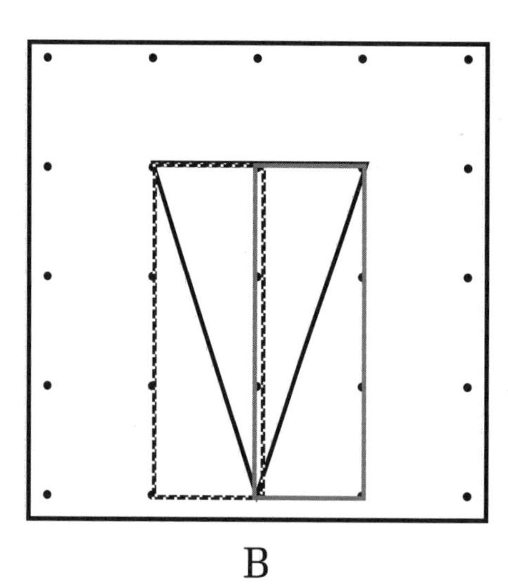

B

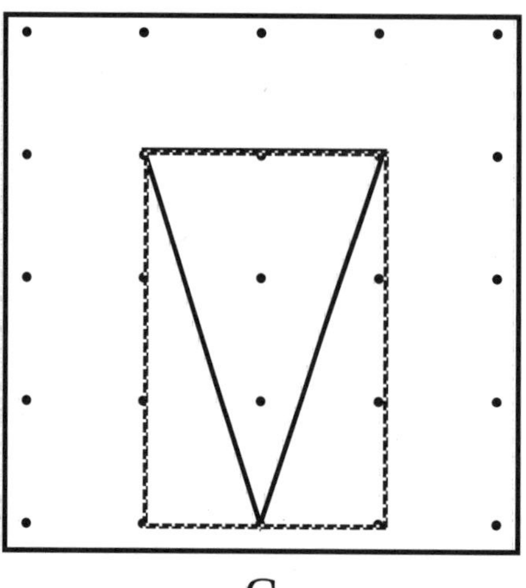

C

# SCALENE TRIANGLES

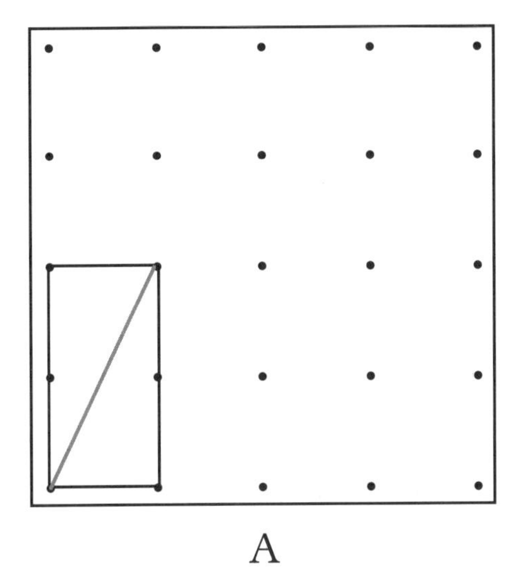

A

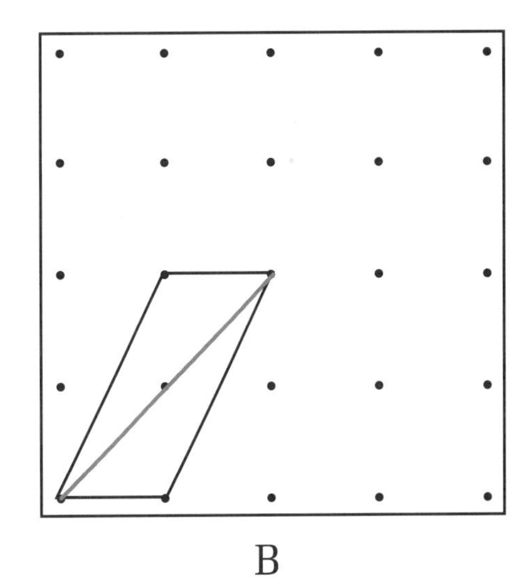

B

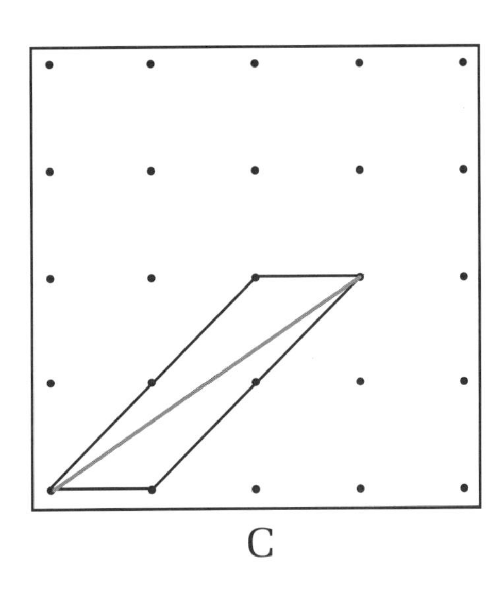

C

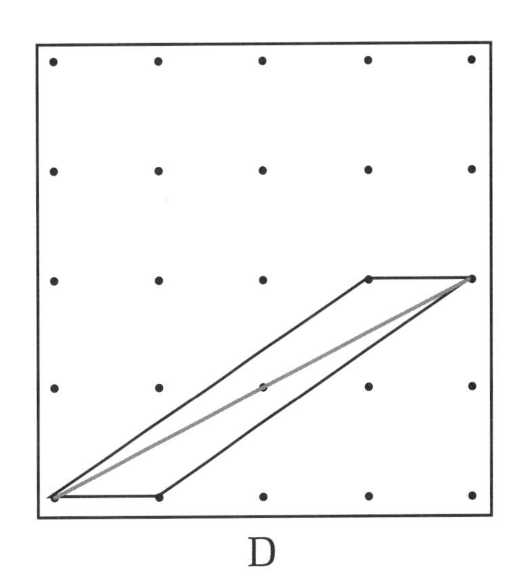

D

## Finding the Area of a Circle

Divide the circle into equal parts:

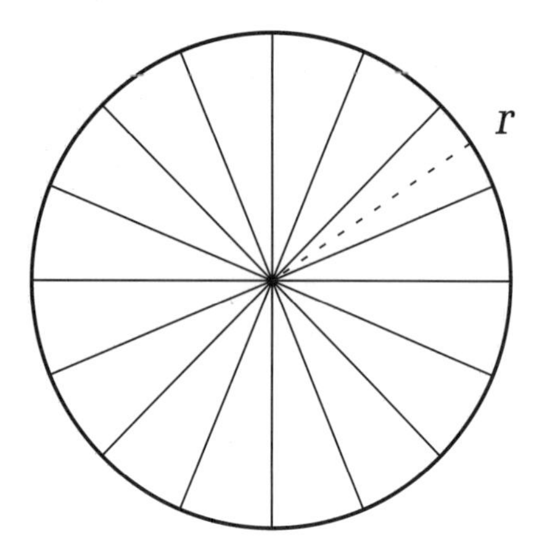


Fit the parts together to form a parallelogram shape:

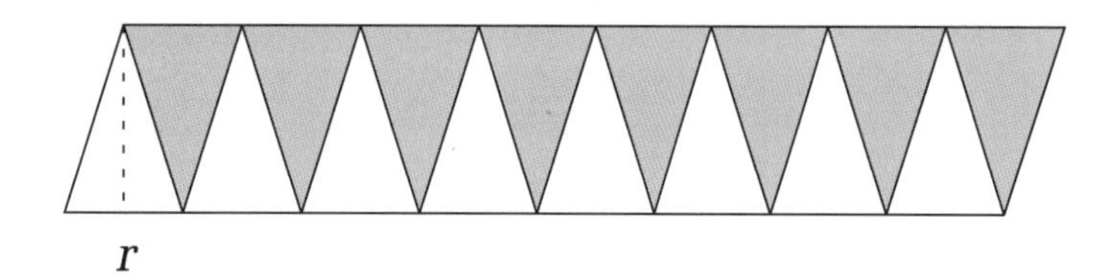


This figure has the same area as the circle.

The base is equal to half the circumference:

$\text{base} = \frac{1}{2} \times 2r$

$\text{base} = r$

The height is equal to the radius:

$\text{height} = r$

The area is approximately equal to the base times the height:

$A = \pi r \times r$

$A = \pi r^2$